# EXPLORING SCIENCE IN THE LIBRARY

## Resources and Activities for Young People

Edited by
Maria Sosa
and
Tracy Gath

**American Library Association**
Chicago and London
2000

Cover by Tessing Design

Text design by Dianne M. Rooney

Composition by the dotted i in Sabon and ITC Kabel using QuarkXPress 3.32 on a Macintosh

Printed on 50-pound white offset, a pH-neutral stock, and bound in 10-point cover stock by McNaughton & Gunn



The paper used in this publication meets the minimum requirements of American National Standard for Information Sciences—Permanence of Paper for Printed Library Materials, ANSI Z39.48-1992. ♾

**Library of Congress Cataloging-in-Publication Data**

Exploring science in the library : resources and activities for young people / edited by Maria Sosa and Tracy Gath.
p. cm.
Includes index.
ISBN 0-8389-0768-7
1. Elementary school libraries—United States. 2. Libraries—United States—Special collections—Science. 3. Science—Study and teaching (Elementary)—United States. I. Sosa, Maria. II. Gath, Tracy.
Z675.S3E97 1999
027.8′222—dc21 99-41496



Printed in the United States of America.

04 03 02 01 00 5 4 3 2 1

# CONTENTS

# FOREWORD

Those of us who were trained in science tend to see science everywhere. We are usually surprised that others don't see the world of science as we do and disappointed that science is not fully integrated into our general culture. For example, a flick of a switch produces light in our living rooms, but the physics that creates electricity remains elusive to far too many people. So it is with the Internet—most users take its presence for granted, not understanding the complex blend of science, mathematics, and technology that has fostered its widespread availability.

Movies and television often reflect our popular culture, but the contributions of scientists and engineers rarely appear as themes. Science-related books are not major components of English classes, although expository materials comprise the majority of reading required in most work settings.

Librarians can help make science more prominent in our culture by including science among the materials made available to children and by exposing children to scientific themes and ideas through a variety of media, including, and especially, good science tradebooks.

My own experience has led me to believe that early exposure to science will produce an appreciation that carries over into adulthood. I am probably involved in science today because of Sputnik. The launch of this satellite by the then Soviet Union sparked a massive reform of American science and mathematics education. Curricula were updated and made more rigorous. Teachers received extensive professional development opportunities. More science and nature books for children were published. The goal was to produce more scientists and engineers, and the clarion call was heard across the United States—even in my segregated, resource-poor elementary school in Birmingham, Alabama. But of all these resources, none was more important to me than books and libraries.

In the late 1950s, Birmingham had only two commercial television stations and the talking heads of early educational television; neither provided real science coverage. Radio broadcasts were limited to music and occasional dramas. Thus, books became my source of adventure and information; they opened up the world beyond Birmingham.

Through books, I was able to meet exotic animals without going to the local zoo, which limited the days Blacks were allowed to visit. And if the nature of the planets or stars wasn't part of my classroom experience, books made it part of my world.

From my earliest memory, books were always very important to me. At first, I bought them, saving every penny I could get my hands on, instead of getting them from the library. In the Birmingham of my childhood, libraries were segregated, too, and not always accessible to Blacks. I was around ten years old before I found a library that welcomed me, but it was a car ride away. Fortunately, the librarian there gave me special permission to take out ten to twelve books at a time, and I made good use of the privilege. Books about ideas and science topics helped me build my vocabulary and reasoning skills. They linked me to the ways in which knowledge and research can be used to solve problems and gain perspective.

The more limited options for schooling that I faced as a child placed a great deal of responsibility on me as an individual. I appreciate how that responsibility forced me to become an independent learner at such a young age. I also believe that a more focused and coordinated set of experiences might have enabled me to come to deeper understandings of the power of scientific ideas and thinking at an earlier age.

No singular event like Sputnik has since captured our imaginations and spurred us toward a collective will to improve education in science, mathematics, and technology. In recent years, a lack of resources has forced many communities, especially in urban centers and rural areas, to reduce access to libraries or limit collections. Although I was fortunate to have a home environment that supported my voracious appetite for learning, many children do not have homes where learning is supported and where books are a regular part of their environment. But rather than focus on what is not available, we can turn our attention to the resources that we do have to open up opportunities for young children. An upturn in the economy has prompted some communities to increase school and library budgets. We have wonderful books, videos, software, and other resources. We have librarians who have chosen this job because of their love of books and their desire to share this love with others. We have libraries with opportunities to expand their roles, opening new vistas for children even where their circumstances may be limited. To see to the stars and beyond—that is the journey we must enable.

SHIRLEY M. MALCOM
Head, Directorate for Education and Human Resources Programs
American Association for the Advancement of Science

# INTRODUCTION

In the past decade, important national efforts to improve science literacy and science education have been launched. These include the American Association for the Advancement of Science's Project 2061 as well as the *National Science Education Standards* of the National Academy of Science. Through Goals 2000 and similar reports, the public has become aware of the importance of keeping our national science standards high in order to maintain our status as a world-class economic power. Scientific and medical breakthroughs, such as the Mission to Mars and the Human Genome Project, have captured the public's attention. All this has increased the demand for science materials in libraries. Although schools and local governments have decreased funding for libraries during the past two decades, an expanding economy is now encouraging many communities to increase school and library budgets. In this shifting climate, the American Association for the Advancement of Science (AAAS) has focused on helping librarians leverage resources to accommodate a growing demand for science materials. Central to this focus are two AAAS projects: the AAAS Science Library Institute and *Science Books and Films* (SB&F). We have drawn on the resources of these two projects to create this book.

*Exploring Science in the Library: Resources and Activities for Young People* is intended to help public and school librarians develop and enhance their skills in science, mathematics, and technology. Much of the material in this book is drawn from an earlier volume, *Great Explorations: Discovering Science in the Library,* which was developed as a training manual for the AAAS Science Library Institute. The Institute sought to make libraries a focal point of school science, mathematics, and technology (SMT) reform by combining theoretical training in education reform issues with practical training using hands-on science and mathematics activities. For more informaiton about the Institute, see the Appendix of this volume.

*Exploring Science in the Library: Resources and Activities for Young People* updates the content of *Great Explorations* and provides new information that examines the role of the librarian in achieving high-quality, standards-based science education for our nation's youth. The book also suggests new directions for the future. An overview of the individual chapters follows.

In chapter 1, "Science, Mathematics, and Technology (SMT) Education Reform," Jerry Bell, director of science, mathematics, and technology education programs at AAAS, reviews five major concepts central to science education reform:

*constructivism,* in which science learning builds on meaning that children construct from all their experiences; this concept is similar to *inquiry-based learning,* which is discussed in chapter 6;

*science for all,* which stresses the goal of science literacy for all students;

*less is more,* which deemphasizes the use of technical jargon and content coverage and emphasizes deeper concept understanding;

*assessment* (also called *authentic assessment*), which allows children to demonstrate their best work and chronicles their progress; and

*systemic reform,* which states that curricula developers, policy makers, and teachers must all be involved in reforming science education.

In chapter 2, "Selecting Science Books for Children," *SB&F* editor-in-chief Maria Sosa discusses the issues involved in selecting effective science tradebooks that support educational reform and address the needs of all students. She also explores the relationship between reading and science and provides concrete suggestions to help librarians select the very best books for their collections. The chapter includes a bibliography of the best children's science books reviewed in *SB&F* during 1998.

In chapter 3, "Meeting the Standards with K–8 Science Tradebooks," *SB&F* contributors Terrence E. Young Jr. and Coleen Salley discuss how librarians can encourage the use of children's literature to integrate science reading across the curriculum. They provide an annotated bibliography of seventy-five science tradebooks that correlate with the K–8 science content standards.

In chapter 4, "Selecting Excellent Science Media," *SB&F Online* editor Tracy Gath explains the process by which *SB&F* reviewers evaluate video and software materials. She provides insight into ways that librarians can select high-quality, effective resources that support standards-based science learning. The chapter includes a list of exemplary resources selected from those reviewed in *SB&F.*

In chapter 5, "Using the Internet to Develop Science Literacy," Maria Sosa suggests ways in which librarians can effectively utilize Internet content that supports the developing national standards for science, technology, and mathematics content and inquiry. She lists over 100 web sites that children and young people can use to support their science exploration, as well as websites that will help librarians enhance their own science and technology literacy skills.

In chapter 6, "Inquiry-Based Learning in the Library," Sosa discusses ways in which librarians can augment science instruction for students by providing informal science activities in the library. These hands-on activities are not intended to replace science instruction in the classroom, but rather to provide additional opportunities for children to explore and experience science. Sosa reviews the basic principles of inquiry-based learning, provides a few sample activities, and lists additional books and Internet sites that support inquiry-based learning. Note that the two essential ingredients of constructivism (discussed in chapter 1)—a student's personal experiences and science activities that help explain those experiences—are also the basis of inquiry-based learning.

Although science programs in the library can be quite inexpensive, even the smallest expense can be a barrier for many libraries. Thus, Maria Sosa, in chapter 7, "Fund-Raising for Science Activities in the Library," offers information about grant writing and about planning, justifying, and conducting a library-based science project. Included are addresses and guidelines for private and corporate foundations and government funders as well as strategies for writing effective grant proposals.

Maria Sosa provides information about and guidelines for forming partnerships in chapter 8, "Partnerships to Promote Science Activities in the Library." We believe that it is critical for librarians to form partnerships and become key players in larger education reform movements. This not only serves the self-interest of librarians and media specialists, it also informs and enriches the education reform community with the unique experiences and perspectives of library professionals.

Finally, in chapter 9, "Science Project Ideas for Libraries and School Media Centers," Mary Chobot, Barbara Holton, and Maria Sosa provide summaries of forty-seven science projects conducted by librarians throughout the country. The authors describe projects devised by librarians in the AAAS Science Library Institutes in Washington, D.C., and Rapid City, South Dakota, as well as exemplary library projects compiled by librarian Barbara Holton for the U.S. Department of Education. This chapter brings together the content of the previous chapters by demonstrating how librarians used their expertise to create partnerships, leverage funds, and develop projects that enhanced the science literacy of the children they served. We hope that sharing these wonderful projects inspires you to explore science with young people at your own library.

# 1

# Science, Mathematics, and Technology (SMT) Education Reform

JERRY BELL

## The Ghost of Science Reform Past

Today, we are realizing that students use their own experiences as a means of learning—in essence, they instruct themselves about concepts. This understanding forms the basis of constructivism, a central tenet in the current educational reform movement. Most of us, however, have already lived through one or more waves of educational reform and may be rather jaded by now. Therefore, it is important to look at how this current movement differs from and builds on those of the past. For example, reforms of a few years ago emphasized that science and mathematics are everywhere and sought to make them "relevant" through such problems as this: Suppose you are having a small party for two of your friends and you want to make pizza for the group. You have a recipe for an 8-inch pizza, but you don't think this will be big enough for three people, so you want to make a 10-inch pizza. This means increasing the amount of each ingredient. The 8-inch recipe calls for 2 cups of flour. How much flour will you need to make a 10-inch pizza?

No matter how you do it, an activity like this leads to a predetermined result along the following lines. Assuming that the larger pizza is made the same thickness as the smaller one, we need to know the relative areas of the circular pizzas in order to figure out how much more flour to use for the larger one. We remember that the area of a circle is given by the product of the square of its radius ($r$) times pi ($\pi$—equal to about 3.1416). The radius of the 8-inch (diameter) pizza is 4 inches and the radius of the 10-inch pizza is 5 inches. Therefore, the areas are $4 \times 4 \times \pi = 16 \times \pi$ square inches and $5 \times 5 \times \pi = 25 \times \pi$ square inches, respectively, for the 8-inch and 10-inch pizzas. We can see that the relative area of the two pizzas is 25 to 16 (about 3 to 2), so it will take about one-and-one-half times as much flour (3 cups) to make the larger pizza.

Although it overstates the case somewhat, we might say that the curricula and activities from past reforms were focused on processes that would lead to a predetermined result. The emphasis was on students carrying out activities that were directed by teachers. The learners' roles were relatively circumscribed, and their work was supposed to lead each of them to the same final correct understanding.

## Science Reform Present—Constructivism

One foundation of the current reforms is *constructivism,* a concept that has grown out of the work of cognitive psychologists over the past few decades. In simplest terms, constructivism suggests that a person constructs meaning from each new experience or piece of information based on previous experiences; there are no blank slates. The educational consequences have been to stress learning activities that are experiential, hands-on, and, to as large an extent as possible, devised by the learner. This provides experiences that are "personalized," because they are actually carried out by the individual who devised them and are therefore likely to overlap with other experiences to facilitate the construction of new meaning. Take the following activity, for example: Set drinking glasses or glass bottles on a table. Cylindrical glasses or bottles of almost any kind, such as soft-drink containers, will work well. The purpose of the activity is to guess first whether the circumference (girth) of the glass is greater or smaller than the height of the glass and then to work in groups to "prove" the guess.

In "proving" their guesses, participants are likely to use several different approaches, many of which amount to comparing the length of paper or string required to go all the way around the glass to the length required to measure the height of the glass. The surprising result of this problem-solving activity is that the circumference is often quite a bit larger than the height, although this is not obvious upon visual examination of the glass. In workshops in which we conduct this activity, everyone seems comfortable with this somewhat counterintuitive finding because each person has participated in deciding how to solve the problem and then has actually carried out the procedure. Hands-on "discovery" (based on methods that come from experience) that the circumference of a cylinder is quite large (more than three times its diameter) has more impact than being told or simply memorizing that the circumference is the product of the diameter times pi.

Constructivism also leads us to consider whether a learning activity or concept is developmentally appropriate for the students with whom we want to use it. If there are no previous experiences upon which to make connections and to build, the concept might be inappropriate. Or it might require an approach that allows students time to develop the necessary experiential base, rather than the efficient, but perhaps ineffective, "tell it like it is" method.

## Paths to Learning

As an analogy, we might imagine that our minds are like forests with a lot of well-worn paths that lead us from one place (piece of information, problem-solving technique, and so forth) to another. As we learn, we are continually increasing the number and extent of the paths and their crossing points (interconnections). If a new idea comes along that is easily accessible from one or more of these paths, we can easily incorporate it into our working knowledge and make connections to it.

However, the new idea might be one that fits best in a remote region of the forest where we have made few paths. Traditionally, educators would give us a map (tell us an idea) and, if we followed the map diligently, we would get to where we needed to be. Unfortunately, this new path and its interconnections were tenuous and fragile.

Constructivism, on the other hand, does not provide a map, but rather asks us to trod our familiar paths (previous experiences)—perhaps several times—to make our own connections to the new idea. Hopefully, the resulting path will be well trodden and easily accessible for future use.

A simple activity that catches students' attention is to investigate the floating and sinking behavior of unopened cans of soft drinks—for example, Coke. Students will likely observe that almost all cans of "regular" Coke sink in water, while almost all cans of diet Coke float. Why? With judicious questioning and discussion, most students will formulate the following logic for themselves: If two objects look identical, but one sinks and one floats in water, the sinker must be heavier than the floater. The regulars must be heavier than the diets. Since the cans are all exactly alike (except for the paint on the label), the difference must be that the content of the regulars is heavier than the content of the diets. The labels tell us that both kinds contain exactly the same volume of beverage, 355 mL, so it must be that the regular liquid is heavier than the diet liquid. This conclusion leads to many further questions and possible investigations, such as: What difference in composition of the beverage could cause the difference in mass?

Communication is critical to "internalizing" learning and to discovering whether what has been constructed has enough connections to be accessible in many different contexts, so that it can be used in solving many different kinds of problems. Students need to "talk through" (and "write through") their understandings and their approaches to problems with peers, teachers, and others. This gives them a chance to find out how well they can use ideas and reinforces connections among ideas (widens and tramples the pathways even further). Libraries and librarians can provide books on hands-on activities that will intrigue students, space and materials for project work, time to tell about it, and resource books and materials for teachers who want to incorporate a constructivist approach in their teaching (see chapter 6).

## Science for All

Demographic data show that the cohort of 4,000,000 students who were high school sophomores in 1977 yielded about 340,000 college freshmen in 1980 who said they were interested in natural sciences, engineering, and mathematics. Of these, about 200,000 graduated with baccalaureate degrees in these fields by 1984, and fewer than 10,000 went on to get Ph.D. degrees in these fields by 1992. The attrition rate is large at each stage of education. This attrition has been called the "pipeline problem." Some believe that too few scientists and engineers are being produced and that this will lead to problems in the future. However, a good deal of controversy exists about whether, in fact, more scientists and engineers will be needed than are getting degrees.

Regardless of the controversy, the overall demographics obscure the fact that women, minorities, and the physically handicapped are underrepresented among those in the "pipeline," and this is a great waste of talent. The numbers, or their usual graphical presentation, that focus on the "pipeline" also divert our attention from the vast majority of students, the rest of the cohort of high school sophomores, who don't express an interest in careers in science, mathematics, and technology.

To address these problems, current reform efforts, in one way or another, espouse the idea that science is for everyone—that is, that excellent science learning opportunities should be available to all, whether or not they have an interest in a science-based career. And these opportunities should be consciously structured to welcome those who have been underrepresented and to encourage them to consider that science is for them as well. Libraries should have science books that illustrate appropriate equity concerns, and librarians should help teachers select and use them.

For example, the American Association for the Advancement of Science's Project 2061 is a science literacy project designed to outline what all 12th grade students should know and understand and to provide guidelines and resources to achieve these goals. Publications available from Project 2061 are the outcome goals in *Science for All Americans,* and the intermediate goals for each grade level in *Benchmarks for Science Literacy,* both of which are based on science for all. These resources as well as other print and computer-based materials, should be available in every library for students, teachers, parents, and others to use. Project 2061 will be discussed in more detail later in this chapter and in chapter 8.

The vision of science and mathematics education that underlies current reform efforts is one of scientific literacy for all students. For example, *Science for All Americans* defines the scientifically literate person as "one who is aware that science, mathematics, and technology are interdependent human enterprises with strengths and limitations; understands key concepts and principles of science; is familiar with the natural world and recognizes both its diversity and unity; and uses scientific knowledge and scientific ways of thinking for individual and social purposes."[1]

Keep in mind, as you refer to the content in Project 2061 and help others to use it, that it represents the science that all students should know and understand, but not all the science that some students (those who really get "turned on" to science) should know and study. For Project 2061 and other science education reform efforts, there is probably no single piece of knowledge or understanding that is absolutely essential. What is essential is the constructivist approach to learning with its emphasis on the teacher as facilitator rather than as information provider.

## Less Is More

One of the catch phrases associated with the present science education reform efforts is "less is more," an obvious oxymoron. What can this mean? Science is not a collection of facts or techniques, yet many textbooks are litanies of facts, new words, and definitions. Current reforms aim to present many fewer facts and concepts and to take the requisite amount of time for students to explore them in some depth in a variety of contexts, so that students have a real working knowledge of several "big ideas," not simply a nodding acquaintance with a lot of words and definitions out of context.

For example, density is a fundamental, but not an easy, "big idea" in terms of explaining many natural phenomena. Developing an understanding of density can be facilitated by a series of activities that build understanding through experience. For example, the teacher can begin with an activity in which students classify small pieces of plastic cut from various household products and containers. Characteristics that students might use for classification include color, texture, transparency, and size. This can be followed by an activity in which students observe the behavior (floating or sinking) of the plastic pieces in three liquids (salt water, fresh water, and rubbing alcohol). Each piece can then be classified according to its floating and sinking behaviors in the three liquids. When students compare their original classifications to the one based on floating and sinking, they will find that some of the previous classifications were better than others at grouping the plastics according to floating and sinking behaviors. The lesson is that some qualities are more useful than others when new criteria or properties are investigated. For students, this activity provides further experience with the property and concept of density of matter. Building a solid understanding through multiple activities of increasing complexity replaces memorizing the formula for density (mass divided by volume), doing a few end-of-chapter problems, and going on to the next topic. Less is more.

Librarians can help teachers by being on the lookout for books and other materials that embody the "less is more" philosophy. These are materials that support depth of learning and will often be characterized by the absence of long lists of technical terms and the presence of multiple perspectives on each topic. In our analogy of the brain and a forest, following a typical textbook is rather like running randomly through the forest and leaving little trace of one's passage. Teachers can do better, and librarians can support, reinforce, and extend their efforts with appropriate resources and library activities.

## Assessment

None of the preceding concepts work unless teachers keep in constant touch with what students are learning. End-of-unit paper-and-pencil tests don't do this very well. They are too late (the unit is over) to help adjust teaching to what is being learned, and they are not usually designed to assess the higher-order thinking and problem-solving skills we want students to develop. Learning has to be assessed on a constant and ongoing basis. We need to assess what is being learned as it is being learned. Assessment can be built into the activities that students do, and teachers can then use the results to determine whether different or additional learning opportunities are needed.

An interesting assessment activity built around the concept of density involves a "Cartesian diver." The device is a capped, 1 liter plastic soft-drink bottle filled to the top with water and containing a glass medicine dropper with a rubber bulb. Holding the bottle in one hand, the presenter commands the "diver" (medicine dropper and bulb) to descend from its starting point at the top of the bottle to the bottom, which it does. Upon the command to rise, the diver again obeys. Cartesian diver devices are then handed out to the students, and they are asked to figure out how to make the diver descend and rise on command. Within a few minutes, most students will discover that gently squeezing the plastic bottle causes the diver to descend and releasing the pressure allows it to rise again. Close examination of the diver reveals that when the bottle is squeezed, the increased pressure in the bottle forces water into the medicine dropper (by compressing the air trapped in it). The extra water in the dropper now is part of the diver and increases its density, so it sinks. When the squeezing stops, the pressure in the bottle decreases, and the extra water in the dropper is forced out by the air trapped inside it. Students who work through this logic for themselves show a high level of understanding of the concept of density. Librarians can facilitate this process by providing resources not only to students who want to assess their own learning but also to teachers who are looking for assessment techniques.

*Authentic assessment* is another of the many catch phrases often used in current reform discussions. In general terms, an assessment is "authentic" if it is based on a real-life problem or situation of interest to the students, if it is related directly to the core of what is being learned (not some peripheral topic), and if it involves students in actually doing things that show their level of understanding. To find activities that can be the basis for authentic assessment, educators can take advantage of the questions and problems that students raise and want to explore, because these are the topics that have already captured their interest.

The hard part is figuring out what to look for as indicators of what the students are learning. Designing scoring systems that will not only accurately record achievement but also be understandable to all stakeholders (teachers, students, administrators, parents, and so on) is a challenge. Libraries should contain resources (for example, journals, books, and copies of articles) for teachers who are trying to develop authentic assessments. Librarians can also provide activities and puzzles for students to use to assess their own learning.

Ideally, assessment of all kinds should be self-assessment. Ongoing and embedded assessment activities should be aimed at helping students become self-assessors as well as at providing information to the teacher. One of the more popular alternative forms of assessment is the use of portfolios. Like artists' portfolios, student portfolios are collections of "best work" that demonstrate growth over time. They are not just a hodgepodge of papers and projects, but a thoughtful collection that demonstrates the student's awareness of his or her own progress and achievement. A librarian can help students select their best work by, for example, acting as a sounding board for students to justify their choices and clarify their rationales.

## The System Approach

In the real world, each part (social, economic, political, and so on) affects many others and is in turn affected by them. The parts are connected in networks of interactions, or "systems." Although essentially everything is connected to everything else through these systems, we often focus our attention on the relatively tightly connected parts of one particular aspect of society, such as education.

At a minimum, the parts of the educational system include students, teachers, librarians, the curriculum, assessment methods, school administrators, parents, the community (local, state, and national), and school governance. As a librarian, you are a part of the system and are probably very much aware of the effects the other parts have on you.

Obviously, a brief overview like this can do no more than provide a glimpse of the richness of present-day science education reform and its various "system parts," but we hope to whet your appetite so you will want to examine at least some of its aspects in depth (to create in your own brain forest well-trod pathways among the many ideas and practices).

Major national and statewide efforts are being made to improve science literacy in the schools. Forty-six states have adopted science standards or curriculum frameworks, many of which are based on standards set by the American Association for the Advancement of Science (AAAS) and the National Academy of Science (NAS). Three reports in particular represent current thinking on content standards in science: the *National Science Education Standards* (National Research Council, 1994); Project 2061's *Benchmarks for Science Literacy* (1993); and the National Science Teachers Association's *Scope, Sequence, and Coordination of Secondary School Science: The Content Core.*[2] The following sections provide brief overviews of the projects responsible for these reports. We also include a bibliography to help you begin your own search and to suggest resources you might wish to make available to teachers, administrators, and parents in your library.

## Project 2061

Project 2061 began as Halley's comet was flying by in 1985. With the help of educators nationwide, the project aims to make everyone science literate by the time the comet returns in 2061. With panels of scientists, mathematicians, and technologists, the project set out to specify what science literacy means. The panels prepared reports that were integrated into *Science for All Americans,* a groundbreaking report that outlined what all high school graduates should know and be able to do in science, math, and technology and that laid out some principles for effective learning and teaching.

To make sure that its work was grounded in the frontline experience of teaching, Project 2061 formed partnerships with six geographically and demographically diverse school district teams of K–12 teachers and administrators. In 1990, the teams set to work designing curriculum models based on *Science for All Americans.* This work became the basis for *Benchmarks for Science Literacy,* published in 1993. *Benchmarks* expands the science literacy goals of *Science for All Americans* into specific learning goals for the end of grades 2, 5, 8, and 12. *Benchmarks* provides sequences of specific learning goals that educators can organize however they choose in designing a core curriculum—one that makes sense to them and meets the goals for science literacy recommended in *Science for All Americans.*

Project 2061 also works to arm teachers with the knowledge and skills they need to help their students work toward science literacy. The project's first CD-ROM, *Resources for Science Literacy: Professional Development,* offers teachers an array of materials to improve their understanding of science literacy and of what it requires of them and their students.

Project 2061 continues to develop new tools that educators can use to change the way they think about and make use of curriculum materials, instructional strategies, and assessments. *Blueprints for Reform*—concept papers on twelve aspects of the education system that need to change to accommodate curriculum reform—is now available both online and in print. *Designs for Science Literacy* will help educators apply design principles as they build K–12 curricula around science literacy goals. The project is also developing the *Atlas of Science Literacy,* a collection of growth-of-understanding maps that graphically show educators the knowledge and skills that students need in order to achieve particular science literacy goals.

## National Science Education Standards

Americans agree that our young people urgently need better science education. But what should students be expected to know and be able to do? Can the same expectations be applied across our diverse society? These and other fundamental issues are addressed in the *National Science Education Standards,* a landmark development effort that reflects the contributions of thousands of teachers, scientists, science educators, and other experts across the country. Coordinated by the National

Committee on Science Education Standards and Assessment of the National Research Council, the *National Science Education Standards* offer a coherent vision of what it means to be scientifically literate, describing what all students, regardless of background or circumstance, should understand and be able to do at different grade levels in various science categories. The standards address:

- The exemplary practice of science teaching that provides students with experiences that enable them to achieve scientific literacy.
- Criteria for assessing and analyzing students' attainments in science and the learning opportunities that school science programs offer.
- The nature and design of both school and district-wide science programs.
- The support and resources needed for students to learn science.

These standards reflect the principles that learning science is an *inquiry-based* process, that science in schools should reflect the intellectual traditions of contemporary science, and that all Americans have a role in improving science education. The full text of the *National Science Education Standards* can be found at the National Academy Press web site: http://www.nap.edu/readingroom/enter2.cgi?ED.html.

## The National Science Teachers Association Scope, Sequence, and Coordination Project

The Scope, Sequence, and Coordination (SS&C) program of the National Science Teachers Association (NSTA) advocates teaching four natural-science subjects each year for all students in grades 9 through 12: biology, chemistry, earth and space sciences, and physics. The program addresses both science literacy and the preparation of those who are motivated to pursue careers in science-related fields.

In 1996, the association published *NSTA Pathways to the Science Standards*. NSTA describes this series as a "travel guide" to putting the standards into practice in the classroom. In three practical editions, written by teachers for teachers of all grade levels, users will find discussions and possible routes for implementing the standards. The guidebooks include classroom examples and resources accessible on a supplementary CD-ROM.

## Bibliography

### Books

American Association for the Advancement of Science. *Benchmarks for Science Literacy.* New York: Oxford University Pr., 1993.

———. *Resources for Science Literacy: Professional Development.* New York: Oxford University Pr., 1997.

———. *Science for All Americans.* New York: Oxford University Pr., 1989.

Atkin, J. Myron, Julie A. Bianchini, and Nicole I. Holthuis. *The Different Worlds of Project 2061.* Paris: Organisation of Economic Cooperation and Development, 1996.

Blank, Rolf K., and Ellen M. Pechman. *State Curriculum Frameworks in Mathematics and Science: How Are They Changing across the States?* Washington, D.C.: Council of Chief State School Officers, 1995.

Center for Science, Mathematics, and Engineering Education, National Research Council. *Every Child a Scientist: Achieving Scientific Literacy for All.* Washington, D.C.: National Academy Pr., 1998.

———. *Improving Student Learning in Mathematics and Science: The Role of National Standards in State Policy.* Washington, D.C.: National Academy Pr., 1997.

———. *Preparation and Credentialing Consistent with the National Science Education Standards: Report of a Symposium.* Washington, D.C.: National Academy Pr., 1997.

National Center for Education Statistics. *The Condition of Education 1997.* Washington, D.C.: U.S. Department of Education, Office of Educational Research and Improvement (NCES 97-388), 1997.

National Education Goals Panel. *The 1995 National Education Goals Report: Building a Nation of Learners.* Washington, D.C.: National Education Goals Panel, 1995.

———. *The 1996 National Education Goals Report: Building a Nation of Learners.* Washington, D.C.: National Education Goals Panel, 1996.

National Research Council. *National Science Education Standards.* Washington, D.C.: National Academy Pr., 1996.

National Science Teachers Association. *NSTA Pathways—Elementary School Edition.* Washington, D.C.: NSTA, 1996.

———. *NSTA Pathways—High School Edition.* Washington, D.C.: NSTA, 1996.

———. *NSTA Pathways—Middle School Edition.* Washington, D.C.: NSTA, 1996.

Schmidt, William H., Curtis C. McKnight, and Senta A. Raizen. *Splintered Vision: An Investigation of U.S. Mathematics and Science Education.* Norwell, Mass.: Kluwer Academic Publishers, 1997.

Suter, Larry E., ed. *Indicators of Science and Mathematics Education 1995.* Arlington, Va.: National Science Foundation, Directorate for Education and Human Resources, Division of Research, Evaluation, and Communication (NSF 96-52), 1996.

Zucker, Andrew A., Viki M. Young, and John Luczak. *Evaluation of the American Association for the Advancement of Science's Project 2061.* Menlo Park, Calif.: SRI International, 1996.

## Internet Resources

### Developing Educational Standards

http://putwest.boces.org/Standards.html

This website, developed by the Putnam Valley Central Schools, contains an annotated list of educational standards and curriculum frameworks documents. Standards are linked by state and by subject area. Links are provided to information about national as well as international standards. Links are also provided to newspaper and magazine articles as well as to organizations involved in education reform.

### Council of Chief State School Officers

http://www.ccsso.org/standards-assessments.html

The Council of Chief State School Officers' site offers resources on a wide range of educational issues, including this standards and assessments page. Here you can find surveys of state progress on standards and examples of standards and benchmarks in math and science. Many of the documents can be downloaded as Adobe Acrobat PDF files.

**ERIC**

The Educational Resources Information Center (ERIC) is a federally funded national information system. Through its sixteen subject-specific clearinghouses, associated adjunct clearinghouses, and support components, ERIC provides a variety of services and products on a broad range of education-related issues.

**AskERIC**
http://ericir.syr.edu/

This website offers a personalized, Internet-based service providing education information to teachers, librarians, counselors, administrators, parents, and others throughout the United States and the world.

**AskERIC Lesson Plan Collection**
http://ericir.syr.edu/Virtual/Lessons/

More than 1,000 unique lesson plans that have been written and submitted to AskERIC by teachers from all over the United States can be found at this website.

**The Eisenhower National Clearinghouse (ENC)**
http://www.enc.org

Voted one of the top ten technology websites for middle school and above students, teachers, and parents, this site is chock full of math and science information. For example, on a monthly basis, ENC provides the following: (1) highlights of thirteen outstanding mathematics and science Internet sites; (2) short features on innovative mathematics and science educators; (3) articles that focus on new ideas currently being used by teachers; (4) resources for teaching mathematics and science topics ranging from incorporating equity into the classroom to recommended CD-ROMs and databases; (5) valuable links to other mathematics and science Internet sites; and (6) an array of mathematics and science lessons and activities.

**National Science Teachers Association**
http://www.gsh.org/nsta/nses_home.htm

The NSTA website includes a resource called "National Science Education Standards, SS&C, and Relevant Learning Materials." Here you may explore and create curricula using either the NSES Content Areas or the SS&C Subject Areas as your starting point. Descriptions and explanations of the Content Standards follow, along with lists of concepts, laws, and theories and the applicable micro-units that you may download. All materials are available in a portable format (Adobe Acrobat Reader PDF files) that can be downloaded, printed, and used in science classes.

## NOTES

1. *Science for All Americans* (New York: Oxford University Pr., 1991).
2. *Scope, Sequence, and Coordination of Secondary School Science: The Content Core* (Washington, D.C.: National Science Teachers Association, 1993).

# 2

# Selecting Science Books for Children

MARIA SOSA

Educational research points to a reciprocal relationship between science and reading. For example, studies indicate that students who read for pleasure and have access to books also do better in science and math. At the same time, inquiry-based science learning helps children develop language and reading skills and strengthen the logical processes needed for content reading. Educators have also long recognized the motivating aspects of the best children's science books, which feature quality, age-appropriate illustrations and text that rival the artistic and literary qualities of the best children's fiction books.

Most librarians, however, are well aware that a flashy exterior is not necessarily an indication of quality. Still, many librarians are more comfortable selecting biographies, poetry, or fiction than science materials. Also, libraries must continue to be vigilant in the administration of funds in spite of a booming economy.

Librarians are especially sensitive to the responsibility they have for assembling a collection of books that are appealing, useful, and free of obvious errors and distortions. Although adults may possess prior knowledge or have a context from which to evaluate the information and perspective provided in a book, this is not likely to be the case with children, especially with technical materials. This chapter attempts to help you enhance your skills in selecting excellent science books for children and young adults in the public library or school media center.

## Qualities of a Good Children's Science Book

Books offer young people opportunities for learning beyond the classroom. They guide young people to one of the major ways that they're likely, as adults, to continue their learning. When a child experiences the pleasure a book offers, reading becomes a habit, and a powerful one at that. Many of the same components that make a good children's fiction book make a good children's science book: presentation, organization, language, and quality of illustrations, for example. But science books must do more than just tell a good story; they must also be accurate and present the processes of science in a way that can be understood by a child without misleading oversimplification. Along with this fundamental requirement of scientific accuracy, good children's science books should encourage an interest in science by prompting young readers to want to know more.

Good science books should also have a well-defined scope and be free of jargon. Although gimmicks, characters, and other motivational features can often help make the subject matter more appealing to young readers, the very best children's science books allow the intrinsic appeal of the subject to shine through. By fostering a genuine interest in a subject, good science books motivate children to explore the topic farther.

Children are naturally curious. This makes them, as many have pointed out, natural scientists. For the very young, science begins by asking questions: Why does it rain? What is thunder? How do birds fly? Why can't I fly? As children grow and learn, the questions (if we're lucky) become more focused: Why are dinosaurs extinct? Where does acid rain come from? How do you get energy from the sun?

Unfortunately, not all children continue to ask such questions. What is it that turns some children off to science as they grow older? Do they lose their curiosity? Or do they simply cease to equate curiosity with science? These questions are not easily answered. However, one way to keep children curious about science is to convince them that science can be fun. Another way is to show young boys and girls that science is relevant by relating it to their lives in an active and direct manner. Successful children's science writers, such as Patricia Lauber and Seymour Simon, effectively use both strategies.

## Assessing Your Collection

Since the 1960s, the volume of scientific information has exploded. If we measured scientific literacy by an individual's knowledge relative to all that is known, even many working scientists and researchers would fail. The sheer magnitude of knowledge available today has made it necessary for us to make choices about what is fundamental—what everybody needs to know. Librarians and media specialists, however, needn't be intimidated. Remember that you are an information specialist. Your role is to provide access to resources, not expertise on subject matter.

You can begin to assess your collection's strengths and deficiencies with regard to science materials by considering the following points:

1. What is the overall size of your collection? What is the ratio of science books to books in other subject areas? What is the average age of the materials in your science collection? Remember, in science, more than in any other subject area, there are few classics. Resources can become quickly outdated as new discoveries are made. Short of replacing your entire science collection every year or so, there isn't much that you can do about this. You can recruit volunteer scientists to help you weed out the most outdated materials, but for some libraries this might mean weeding out half the collection. A better idea is to develop a strategic plan to update your collection regularly, focusing on the subject areas that are most lacking.

2. Which science topic is best represented in your collection? Which is least represented? Does this match your needs? In other words, if you have more books on rocks and minerals than on any other science topic, is there a valid reason for this? Perhaps your library has made a decision to specialize in a certain subject area because it has local or regional appeal. For example, a library in Florida may have a large collection of books on hurricanes. This is fine, as long as you review that decision regularly to make sure that it still suits your needs.

3. Do you have quality materials for all ages and levels represented in your school or community? If not, which level seems to be the most deficient? Are there any valid reasons for this deficiency? A valid reason might be that there simply are no quality materials available on a particular subject for a particular age group. You can check Project 2061's *Benchmarks for Science Literacy* to help you determine the especially relevant topics for children at various educational levels.

4. What curriculum areas do you need to support with your collection? Are these well represented? Work with the teachers to consider not only what they are currently teaching, but also what they might like to introduce to their classes if they had the appropriate resources.

5. Does your collection consist of a good mixture of different types of books—that is, hands-on science books, science-related fiction, informational books on single topics, and reference books? It is important that children have access to a wide variety of genres.

After answering these questions, you may well decide that your collection needs to be enhanced. This brings us back to the issue of how best to utilize your financial resources in order to achieve your goal of a balanced, accurate, and appropriate science collection. One thing is certain: To acquire anything short of the best available materials would be a disservice to readers. For help in choosing the best items, you may wish to use expert reviews.

## Finding Expert Reviews

Many tools provide expert reviews. Some publications are devoted entirely to reviewing, such as *Science Books and Films (SB&F),* which publishes an annual Best Books for Children list. The National Science Teachers Association (NSTA) also publishes *Outstanding Science Trade Books for Children* in partnership with the Children's Book Council. This list can be found at the NSTA website at http://www.nsta.org/pubs/sc/ostblist.htm. In addition to review journals, look for expert reviews in specialized publications for educators, such as those published by the NSTA, the National Association of Biology Teachers (NABT), the American Chemical Society (ACS), and others.

When turning to an expert review, the librarian needs to feel confident that the reviewer is indeed an expert. Therefore, look for publications that list the affiliations of each reviewer. Also, look for review sources that include a detailed explanation of the criteria by which materials are evaluated. For example, *SB&F* reviewers are given the following general instructions, which are summarized in each issue:

> Please write a critical evaluation of approximately 200 words (slightly more, if necessary), emphasizing the merits and/or demerits of the book

> (and accompanying supplements, if any). Describe and critique the content, technical quality, and instructional value. Take special note of the overall quality of the presentation of facts, theories, and processes of science and their interrelationships. For example, does the material accurately depict the uses and limitations of the scientific method? Indicate for which audiences the material is most appropriate and why. Also mention how the material could be used (reference, classroom, general awareness, other).

The more expensive the product you are considering purchasing, the more you need to get a second or even a third opinion. Remember that reviewers are human and that in the vast majority of cases, only one person reviews the material.

Don't forget to ask the experts in your own building. If you are a school media specialist, work with your science teachers. Ask the children themselves. Consider having focus groups in which you invite a group of students to the library and examine their attitudes about science books. Find out what works for them and what doesn't. An adult reviewer can praise a book, but if a child is turned off, the target is missed. You can also form a rotating panel of young reviewers who can recommend books for other children.

## The Importance of Factual Accuracy

Scientists formulate and test their explanations of nature using observations, experiments, and theoretical and mathematical models. It is important that students recognize that science is not a collection of static facts and that scientific knowledge is subject to modification as new information challenges prevailing theories and as new theories lead to looking at old observations in new ways. Thus, we want to avoid the impression that science is a collection of facts.

However, some facts are more important than others. For example, one book may say that there are 10 billion nerve cells in the brain while another may say there are 100 billion. Such a discrepancy may arise because new discoveries in neuroscience are so abundant that it is difficult for books to be completely current. Although you may not necessarily want to discard a book for citing the lower number of neurons, you may rightly be wary of a new book that does not reflect of current developments. Keep in mind, however, that there are areas in which scientists legitimately disagree—global warming, for example. The best reason for rejecting a book with a blatant error (for example, that Venus is the closest planet to the Sun) is that such carelessness with regard to readily available facts may carry over to less well known details.

## Avoiding Dangerous Science Books

Can science books be dangerous? Although the question may seem facetious, there are some scenarios in which inappropriate materials can actually be harmful.

1. A boring science book can be dangerous, for such a book conveys to children the sense that science is uninteresting, tedious, or difficult. Books can be boring because they are poorly written or illustrated or, worse, because the author cannot convincingly convey the excitement of science.
2. Another dangerous type of science book is one that makes children feel that science excludes people of their gender, race, or socioeconomic status. Books that use settings or examples that are not relevant to all children can make some readers feel that science is not appropriate for them.
3. Books that portray science as a static collection of facts, rather than a process, endanger children's innate sense of wonder and curiosity about the natural world.
4. Books that are inaccurate to the point of creating or fostering misconceptions are dangerous because such misconceptions can be stumbling blocks to children's understanding of science.
5. Hands-on science books that don't take safety into account can be dangerous for obvious reasons. Some safety issues to consider are detailed below:

In general, when evaluating hands-on science activities, consider whether children will have to handle any of the following: chemicals that are poisonous, harmful to their skin or eyes, or flammable; flames or electrical currents with enough power to shock or burn; or knives, razor blades, broken glass, or other objects that can cut. If an activity includes one or more of these things, you need to make sure that appropriate safety precautions are listed, whether the activity indicates that it should be done with adult supervision, and the age of the children who will do the activity.

## Selecting Books That Open Doors

Books help us to set our own perspectives. Very early on, children move from *learning to read* to *reading to learn.* They shift from learning the mechanics of pronunciation, grammar, syntax, the flow of ideas, and what the author is saying to them to using the books to find out what *they* want to know. Children, as they grow, become selective and begin to search for what interests them.

Reading to learn ought to be the central theme of children's science books. Children are curious; they have many questions about the world around them, and, with guidance, they may learn to ask many different kinds of questions. Tradebooks put science in the context of people, places, and the methods that lead to a discovery or a new theory, stimulating more interest—and more questions—by the reader. Textbooks, on the other hand, typically present facts superficially, ignoring the power of good storytelling to create excitement as well as to inform. Good tradebooks take advantage of a child's personal interest at a time when the child wants to know something. Good tradebooks open children's eyes to things they didn't know—or didn't know they wanted to know.

Good books also let children see the inside of how one comes to know. To simply present a fact without helping a child see how we came to know that fact doesn't promote good science learning. A good book can promote such learning by sharing information about the people behind a scientific enterprise, by discussing the ramifications of knowing, and by giving the reader a glimpse into a scientist's mind. A good book won't tell all the answers. It helps children know how much is known and what else there may be to find out, and it gives them clues to help them find out other things—hands-on activities, other books, and guides.

Good science books help children understand how science is done by going beyond traditional recitations of facts to see how scientists really do their work. The fascinating ways in which scientific work is done; the train of logic that is followed in building and expanding ideas, models, and schema; and the testing of those ideas against evidence can all be incorporated into well-told children's science stories.

It is important that students realize that doing science involves many different kinds of work and engages men and women of all ages and backgrounds. Biographies of scientists can help do this. Books not only can open up worlds of interest, they can put faces on those worlds. For children who might not perceive a future for themselves beyond the limits of their present circumstances, a book can introduce them to people who not only look like them, but who made career choices they may never have considered or even known existed.

## Utilizing Human Resources

Like Rome, a model science, mathematics, and technology library collection cannot be built in a day. Nor can just one person build it. The human resources required to build your collection are as important as the books, videos, and software that line your shelves. The following tips can help you get started.

1. Work with science teachers in your school or community to select and focus on an area that correlates with what they are teaching. After you have built up one area, it will be easier to convince administrators that other areas need to be enhanced as well.

2. Make an effort to involve parents in your library programs. If you don't already have one, consider an inexpensive newsletter that informs parents about the library and about what it needs. Public librarians can enlist parental assistance in enhancing the science offerings in the children's section. School librarians can work with parents to devise fund-raising strategies and to develop family programming that is centered in the library. Once parents understand how an exemplary collection can help their children achieve in science and mathematics, they will be your staunchest supporters.

3. Bring your science resources to the attention of library patrons through displays and special activities. When something good happens in your library, spread the news. Many local newspapers will be happy to feature stories about your efforts to enhance science, mathematics, and technology programs in your library.

4. Start a collection in your library on science and mathematics reform. Keep teachers informed of such materials and encourage them to share reform issues with parents. Teachers may not have time to stay abreast of current information about research and standards-based

reform. You can be a resource for them on exemplary practices and new products for the classroom. Much information is available online through electronic bulletin boards.

5. Develop a resource list in conjunction with local companies, science organizations, university departments, and other community groups, and join them in devising fund-raising and program strategies. These might include sponsoring lectures or readings, forming children's book clubs, and developing hands-on activity kits or materials to check out with the books. This can position libraries to support out-of-school learning in science, increase the interest and involvement of donors and organizations that have a special interest in the sciences, and support the National Education Goals in science and mathematics. A more comprehensive approach might involve teachers and area scientists in a review of current titles and holdings with joint planning of a collection.

You can see how it could take some effort to develop a collection of equitable, excellent science books that meets the needs of all the children you serve. At the same time, you can imagine the possibilities that would open up for you, for teachers and parents, and especially for students as everyone uses these quality books to expand their understanding of the world.

# 3

# Meeting the Content Standards with K–8 Science Tradebooks

TERRENCE E. YOUNG JR.

COLEEN SALLEY

Science literacy is important for many reasons, not the least of which is that all Americans will be called upon to make decisions about issues that concern science. Science and technology form too large a part of our lives to be ignored, and informed citizens will make the best decisions. To prepare children for the responsibilities of citizenship in a technological society, reading and language arts teachers, science teachers, and librarians can work together by using children's literature to integrate science reading across the curriculum. Based on research studies, the national standards in both science and language arts direct elementary schools to integrate science with other subjects, thereby helping students become critical thinkers and decision makers.

Children enjoy science; most have an innate curiosity that seems perfectly suited to learning about the wonders of science. Both beginning and sophisticated readers often choose science as their favorite subject. The picture books for the youngest readers (and nonreaders) provide a wonderful opportunity for shared learning experiences—reading a beautiful and accurate science book to a young child opens up a world of wonder to them. As they grow, it is important to provide children with interesting books that they can read and enjoy on their own, for this is the habit that leads to lifelong reading. Good science books that understand the way children learn have much to contribute. Young readers can gain confidence when they are able to read and answer for themselves some of the questions that have been racing through their minds.

School library media specialists agree with other educators that there is a need for achieving higher levels of science literacy by American students. Our students need better science education. But what is the role of library media specialist in science education? What should we be expected to know and be able to do?

## The National Science Education Standards

The *National Science Education Standards,* published in December 1995, describe exemplary teaching practices that provide students with experiences that lead to scientific literacy.[1] The standards are a complete set of outcomes; they do not prescribe a curriculum.

The standards urge educators to replace traditional teaching methods with stimulating learning experiences that mirror the excitement of the scientific process itself. As students' interest levels increase, so will their levels of learning. The science content standards provide indicators for what students should know, understand, and be able to do from kindergarten through high school. For example, by the eighth grade, students should have a basic understanding of properties and changes of the properties of matter, diversity and adaptations of organisms, the structure of the Earth system, and the use of science and technology in solving simple problems.

The *National Science Education Standards* are divided into six categories. There are standards for (1) science teaching; (2) professional development for teachers of science; (3) assessment in science education; (4) science content; (5) science education programs; and (6) science education systems.

The science content standards outline what students should know, understand, and be able to do in the natural sciences over the course of K–12 education. They are divided into eight categories:

The standard category, Unifying Concepts and Processes in Science, is presented for all grade levels, because the understandings and abilities associated with these concepts need to be developed throughout a student's educational experiences. Unifying Concepts and Processes in Science underlies and complements the following standards:

Content Standard Category
- A—Science as Inquiry
- B—Physical Science
- C—Life Science
- D—Earth and Space Science
- E—Science and Technology
- F—Science in Personal and Social Perspective
- G—History and Nature of Science.

CONTENT STANDARD CATEGORY A

| *Levels K–4* | *Levels 5–8* | *Levels 9–12* |
|---|---|---|
| Abilities necessary to do scientific inquiry | Abilities necessary to do scientific inquiry | Abilities necessary to do scientific inquiry |
| Understanding about scientific inquiry | Understanding about scientific inquiry | Understanding about scientific inquiry |

CONTENT STANDARD CATEGORY B

| *Levels K–4* | *Levels 5–8* | *Levels 9–12* |
|---|---|---|
| Properties of objects and materials | Properties and changes of properties of matter | Structure of atoms |
| Position and motion of objects | Motion and forces | Structure and properties of matter |
| Light, heat, electricity, magnetism | Transfer of energy | Chemical reactions |
| | | Motions and forces |
| | | Conservation of energy and increase in disorder |
| | | Interactions of energy and matter |

CONTENT STANDARD CATEGORY C

| *Levels K–4* | *Levels 5–8* | *Levels 9–12* |
|---|---|---|
| Characteristics of organisms | Structure and function in living systems | The cell |
| Life cycle of organisms | Reproduction and heredity | Molecular basis of heredity |
| Organisms and environments | Regulation and behavior | Biological evolution |
| | Populations and ecosystems | Interdependence of organisms |
| | Diversity and adaptations of organisms | Matter, energy, and organization in living systems |
| | | Behavior of organisms |

## CONTENT STANDARD CATEGORY D

| *Levels K–4* | *Levels 5–8* | *Levels 9–12* |
|---|---|---|
| Properties of earth materials | Structure of the earth system | Energy in the earth system |
| Objects in the sky | Earth's history | Geochemical cycles |
| Changes in earth and sky | Earth in the solar system | Origin and evolution of the earth system |
| | | Origin and evolution of the universe |

## CONTENT STANDARD CATEGORY E

| *Levels K–4* | *Levels 5–8* | *Levels 9–12* |
|---|---|---|
| Abilities to distinguish between natural objects | Abilities of technological design | Abilities of technological design |
| Abilities of technological design | Understanding about science and technology | Understanding about science and technology |
| Understanding about science and technology | | |

## CONTENT STANDARD CATEGORY F

| *Levels K–4* | *Levels 5–8* | *Levels 9–12* |
|---|---|---|
| Personal health | Personal health | Personal and community health |
| Characteristics and change in populations | Populations, resources, and environments | Population growth |
| Types of resources | Natural hazards | Environmental quality |
| Changes in environments | Risks and benefits | Natural and human-induced hazards |
| Science and technology in local challenges | Science and technology in society | Science and technology in local, national, and global challenges |

## CONTENT STANDARD CATEGORY G

| *Levels K–4* | *Levels 5–8* | *Levels 9–12* |
|---|---|---|
| Science as a human endeavor | Science as a human endeavor | Science as a human endeavor |
| | Nature of science | Nature of scientific knowledge |
| | History of science | Historical perspective |

## The National Assessment of Educational Progress

In May 1997, the education community received good news with the release of the science results of the 1996 National Assessment of Educational Progress (NAEP) for eighth grade.[2] The assessment measured students' knowledge of earth, life, and physical science facts as well as students' ability to use those facts along with the tools, procedures, and reasoning processes of science. For more than a quarter of a century, NAEP has reported to policy makers, educators, and the general public on the educational achievement of students in the United States. As the nation's only ongoing survey of students' educational progress, NAEP has become an important resource for information on what students know and can do. The NAEP 1996 science assessment (http://nces.ed.gov/nationsreportcard/science/science.asp) continues the congressional mandate to evaluate and report the educational progress of science students at grades 4, 8, and 12 at least every four years. The national results describe students' science achievement at each grade and within various subgroups of the general population. The NAEP 1996 Science Report Card for the Nation and the States: Executive Summary can be read at http://nces.ed.gov/nationsreportcard/96report/97497.shtml. The NAEP results are considered the most reliable data for comparing the performance of students among states. The positive results from the 1996 science NAEP are the culmination of a strong curriculum, teachers who are well-prepared to teach the curriculum, and school library media specialists who play a role in the effort to help science teachers continually improve their skills.

## Using MARC Records

Curriculum Enhanced MARC (CEMARC) records provide media specialists with specific fields for curricular information. New or existing records can be modified to reflect this information. Reviews (tag 520) and grade, interest, and motivation levels (tag 521, first and second indicators) are found in many commercially available records. The 658 tag may be used for local or national curriculum objectives as well as state objectives or all objectives as the field is repeatable. There is provision for the degree to which the item is correlated—minimum, core, or full. Additional information and examples are available on Follett Software Company's web page at http://www.fsc.follett.com. The "tag of the month" reviews the 658 tag. You can have your own school district's curriculum code designated in USMARC Relators by contacting the Library of Congress. When content standards information is added to catalog records, both school library media specialists and teachers can retrieve relevant books that correlate with teaching objectives and benchmarks. When selecting library materials, school library media specialists should select science materials that support the national science content standards and reinforce concepts on the NAEP science assessment.

The following bibliography is correlated to the *National Science Education Standards* using the format Content Standard, Grade Level, Objective. For example, C, K–4, 1–3 means the item meets Content Standard C, Life Science, the item is suitable for grades K–4, and it meets the following objectives: (1) Characteristics of organisms, (2) Life cycle of organisms, and (3) Organisms and environments.

We hope that you will take some time to familiarize yourself with the *National Science Education Standards* and support them in your classrooms, library programs, and online catalogs. Our own online catalogs are ready to link library resources to learning units, subject interests, and environments. School library media specialists must do their part to ensure that all students not only know, but also understand and enjoy science.

## Bibliography

Aldridge, Bill. *The Ultimate Science Quiz Book*. New York: Watts, 1994. 143p. Grades 5–9. All standards.

Includes index.

Aliki. *My Visit to the Zoo*. New York: HarperCollins, 1997. 32p. Grades PreK–2. C, K–4, 1–3.

A day at the zoo introduces the different animals that exist in the world, where they come from, what their natural habitats are like, whether or not they are endangered, and the role zoos and conservation parks play today.

Arnold, Caroline. *African Animals*. New York: Morrow, 1997. 48p. Grades PreK–1. C, K–4, 1–3.

Describes animals of the African continent and explains how each is able to adapt to its special environment.

———. *Hawk Highway in the Sky: Watching Raptor Migration.* Photographs by Robert Kruidenier. San Diego, Calif.: Harcourt, 1997. 48p. Grades 4–8. C, 5–8, 1–3.

Provides information about hawks, eagles, and falcons. Useful appendix. Index.

Arnosky, Jim. *Crinkleroot's Guide to Knowing Butterflies and Moths.* New York: Simon & Schuster, 1996. 32p. Grades K–3. C, K–4, 1–3.

An illustrated introduction to the appearance and habits of various butterflies and moths.

———. *Watching Water Birds.* Washington, D.C.: National Geographic Society, 1997. 32p. Grades K–3. C, K–4, 1–3.

Personal observations of various species of freshwater and saltwater birds, including loons and grebes, mergansers, mallards, wood ducks, Canada geese, gulls, and herons; illustrated with extraordinary art.

Aronson, Billy. *Eclipses: Nature's Blackouts.* New York: Watts, 1996. 64p. Grades 5–8. D, 5–8, 1 & 3.

Explains what causes eclipses of the sun and moon and describes how they have been viewed and studied at different times. Glossary; Further Reading; Internet Resources; Index.

Artell, Mike. *Starry Skies.* Glenview, Ill.: Good Year, 1997. 96p. Grades 3–7. D, K–4, 1–3.

Interesting text is complemented by colorful cartoon art that entertains as well as instructs about the universe.

Asch, Frank. *Cactus Poems.* Photographs by Ted Levin. San Diego, Calif.: Harcourt, 1998. 48p. All grades. C, K–4, 1–3.

Beautiful photos and descriptive poems celebrate the odd and awesome aspects of the North American deserts—the Sonoran, the Mojave, the Great Basin, and the Chihuahuam. Notes on Desert Life in appendix.

———. *Sawgrass Poems: A View of the Everglades.* Photographs by Ted Levin. San Diego, Calif.: Harcourt, 1996. unp. All ages. C, K–4, 3; C, 5–8, 4–5.

Poems and photographs provide a picture of the unique system of the Florida Everglades. Notes; photo captions.

Bang, Molly. *Chattanooga Sludge.* San Diego, Calif.: Harcourt, 1996. 42p. Grades 2–7. E, K–4, 3; E, 5–8, 1 & 2.

Although the format may lead readers to expect a story for preschoolers, this unusual picture book describes a scientist's experimental use of microbes to turn toxic sludge into clear water. Environmentally friendly.

Bare, Colleen. *Never Kiss an Alligator!* New York: Dutton, 1989. 30p. Grades K–3. C, K–4, 1.

Easy-reading text, great photos. Discusses in simple language the characteristics and habits of alligators.

Barner, Bob. *Dem Bones.* San Francisco, Calif.: Chronicle, 1996. 26p. Grades PreK–K. C, K–4, 1.

An introduction to human anatomy that is a rollicking read-aloud, sing-along treat for children as they learn anatomy, rhyme, and language. "Dem Bones" is a well-known African American spiritual. Scientific facts and names combined with lyrics make this a fascinating book.

Bell, Karen Magnuson. *Fire in Their Eyes: Wildfires and the People Who Fight Them.* San Diego, Calif.: Harcourt, 1999. 64p. Grades 4–up. F, 5–8, 3–4.

Depicts in text and photographs the training, equipment, and real-life experiences not only of people who risk their lives to battle wildfires, but also of people who use fire for ecological reasons.

Bishop, Nick. *The Secrets of Animal Flight.* Boston: Houghton, 1997. 32p. Grades 3–6. C, 5–8, 1.

Photos of birds, bats, and insects in flight. Clear explanations with photographs and diagrams demonstrating the process. Bibliography.

Blake, Robert J. *Akiak.* New York: Philomel, 1997. 32p. Grades K–3. C, K–4, 3.

Introduces young readers to the Iditarod through the courageous journey of a lone dog. Akiak the sled dog refuses to give up after being injured during the Iditarod race.

Bosveld, Jane. *While a Tree Was Growing.* New York: Workman, 1997. 48p. Grades 3–8. C, 5–8, 1–5.

Juxtaposes the life of a single living sequoia with a chronology of world events to illustrate 3,500 years of history, natural science, anthropology, and more while telling the story of the tree's life. An American Museum of Natural History book.

Brown, Philippa-Alys. *Kangaroos Have Joeys.* New York: Atheneum, 1996. unp. Grades K–3. C, K–4, 1.

Rhyming text and vivid art introduce the often unusual names of animal offspring. Includes a detailed appendix. Also *A Gaggle of Geese,* a read-aloud science book.

Brown, Ruth. *Toad.* New York: Dutton, 1996. unp. Grades K–3. C, K–4, 1.

Gruesome words and pictures will delight young readers. Great for descriptive adjectives in language arts class.

Butterfield, Moira. *What Am I?* Series. Illustrated by Wayne Ford. Austin, Tex.: Raintree Steck-Vaughn, 1998. 32p. Grades PreK–3. C, K–4, 1–3.

This series, in the form of a riddle, is a lot of fun for young children and beginning readers. Excellent illustrations add to the fun. A science read-aloud series. Titles are:

*Fierce, Strong, and Snappy* (alligator)

*Big, Rough, and Wrinkly* (elephant)

*Bouncy, Big, and Furry* (kangaroo)

*Brown, Fierce, and Furry* (bear)

*Fast, Strong, and Striped* (Bengal tiger)

*Jumpy, Green, and Croaky* (frog)

*Quick, Quiet, and Feathered* (owl)

Bryant-Mole, Karen. *Picture This: Seasons*. Series. Des Plaines, Ill.: Heinemann, 1997. 24p. Grades PreK–2. D, K–4, 3.

The seasons are introduced with labeled photographs that identify things associated with each season: flowers, weather, clothes, the garden, holidays, vegetables, and so on. Titles are:

*Autumn*
*Summer*
*Spring*
*Winter*

Bunting, Eve. *Secret Place*. Illustrated by Ted Rand. New York: Clarion, 1996. 26p. Grades K–3. F, K–4, 2.

If you look carefully, nature can be seen in the heart of the city.

Calhoun, Mary. *Flood*. Illustrated by Erick Ingraham. New York: Morrow, 1997. 40p. Grades K–2. D, K–4, 3.

One fictional Midwest family is forced to leave their home during the flooding of the Mississippi River in 1993.

Cannon, Janell. *Verdi*. San Diego, Calif.: Harcourt, 1997. 46p. Grades PreK–3. C, K–4, 1–3.

A young python does not want to grow slow and boring like the older snakes he sees in the tropical jungle in which he lives.

Chandler, Gary, and Kevin Graham. *Making a Better World*. Series. Brookfield, Conn.: Twenty-First Century Books, 1996. 64p. Grades 5–8. A, 5–8, 1–2; E, 5–8, 1–2; F, 5–8, 2–5.

Presents examples of successful efforts to protect natural resources and wildlife. Includes the names and addresses of organizations that are involved in these endeavors. Titles are:

*Recycling*
*Guardians of Wildlife*
*Protecting Our Air, Land, and Water*
*Alternative Energy Sources*
*Kids Who Make a Difference*
*Natural Foods and Products*

Cherry, Lynne. *Flute's Journey: The Life of a Wood Thrush*. San Diego, Calif.: Harcourt, 1997. unp. Grades 2–5. C, K–4, 2.

Follows the journey of a young thrush from Maryland to Costa Rica and back again.

Clark, Margaret Goff. *Save the Florida Key Deer.* New York: Cobblehill, 1998. 38p. Grades 3–6. C, 5–8, 1–5.

Discusses the history, physical characteristics, behavior, and habitat of the small "toy" deer that have lived for hundreds of years in the Florida Keys, as well as threats to their continued existence.

Cole, Henry. *Jack's Garden.* New York: Greenwillow, 1995. 26p. Grades 2–5. C, K–4, 2.

Predictable book. Following the rhythmic cadence of "This Is the House That Jack Built," Henry Cole's cumulative text and illustrations depict what happens in Jack's garden after he plants his seeds.

Collard, Sneed, III. *Animal Dads.* Illustrated by Steve Jenkins. Boston: Houghton, 1997. 32p. Preschool–2. C, K–4, 1–3.

Illustrations and simple text describe how the males of different species help take care of their young.

Couper, Heather, and Nigel Henbest. *Big Bang: The Story of the Universe.* New York: Dorling Kindersley, 1997. 45p. Grades 4–12. D, 9–12, 1–4.

This book follows the story of the Universe from its birth to the present—and beyond. Each successive spread relates the next stage in the unfolding saga, so it is best to read the book in sequence. Glossary; Index.

Cowcher, Helen. *Jaguar.* New York: Scholastic, 1997. 32p. Grades K–5. C, K–4, 1–3.

Fabulous art supports a mystical mood in this fifth title promoting preservation of our world. Great for discussion. There is a "More about Jaguar" section at the back of the book. A read-aloud book.

Cumbaa, Stephen. *The Bones Book and Skeleton.* New York: Workman, 1991. 64p. Grades 4–8. C, 5–8, 1.

Filled with projects, experiments, and incredible facts, this book will satisfy a young anatomist's guide to the body. Includes a skeletal model. Companion books: *The Neanderthal Book and Skeleton,* 1997; *The Bones and Skeleton Game Book,* 1993.

Darling, Kathy. *Komodo Dragon.* Photographs by Tar Darling. *On Location.* Series. New York: Lothrop, 1997. 40p. Grades 4–8. C, 5–8, 1–2.

Describes the physical characteristics, habitat, and behavior of the giant lizards found only in Indonesia. Index; Dragon Facts. Also in series: *Chameleons.*

Davies, Nicola. *Big Blue Whale.* Illustrated by Nick Maland. Cambridge, Mass.: Candlewick, 1997. 30p. Grades PreK–2. C, K–4, 1–3.

Details about the biggest creature that has ever lived written in a flowing style with beautiful illustrations.

Davis, Susan. *The Sporting Life: Discover the Unexpected Science behind Your Favorite Sports and Games.* New York: Henry Holt, 1997. 151p. Grades 4–9. F, 5–8, 5. All standards.

Includes Bibliography; Index.

Dewey, Jennifer Owings. *Poison Dart Frogs.* Homesdale, Pa.: Boyd Mills Pr., 1998. 32p. Grades 2–5. C, K–4, 1–3.

A variety of colorful and tiny poison dart frogs living in the rain forest of Central and South America are pictured in their natural habitat. Topics include mating habits, natural predators, methods to extract the frog's poison, and its unique nurturing habits. Colorful illustrations.

Dunn, Andrew. *The Children's Atlas of Scientific Discoveries and Inventions.* Brookfield, Conn.: Millbrook, 1997. 96p. Grades 3–6. All standards.

The earliest discoveries of humans and the flights of imagination that brought about the most technologically advanced inventions of the twentieth century.

Dunphy, Madeleine. *Here Is the Wetland.* New York: Hyperion, 1996. unp. Grades PreK–2. C, K–4, 1–3.

Lyrical prose provides a clear understanding of how each living creature is linked to the others in a vital chain of life. Uses a cumulative approach to describe the wetland ecology of a freshwater marsh, the most common type of wetland in North America.

Earle, Syvia. *Hello, Fish! Visiting the Coral Reef.* Photographs by Wolcott Henry. Washington, D.C.: National Geographic Society, 1999. 32p. Grades K–3. C, K–4, 1–3.

An underwater explorer takes a tour of the ocean and introduces such fish as the damselfish, stargazer, and brown goby. Stunning photography brings the text to life.

Edmonds, Alex. *Closer Look.* Series. Brookfield, Conn.: Copper Beech Books, 1997. 32p. Grades 4–6. E, K–4, 1 & 3; F, K–4, 4 & 5.

This fascinating new series is designed to appeal to the environmentally aware generation. It brings a fresh approach to the presentation of environmental issues for the young reader of the twenty-first century, and combines clear, informative text with maps, diagrams, and photographs. Titles are:

*Acid Rain*
*The Ozone Hole*

Esbensen, Barbara. *Swift as the Wind.* Illustrated by Jean Cassels. New York: Orchard, 1996. unp. Grades K–3. C, K–4, 1.

Physical characteristics and behavior of the cheetah, illustrated with remarkable art.

Fleisher, Paul. *Life Cycles of a Dozen Diverse Creatures.* Brookfield, Conn.: Millbrook, 1996. 80p. Grades 3–6. C, K–4, 2.

Compares and contrasts the life cycles of twelve animals, including the microscopic daphnia, seahorse, oyster, honeybee, and blood fluke. Explains the concept of the life cycle. Includes scientific names, glossary, further reading, and index.

Fleming, Denise. *Where Once There Was a Wood.* New York: Henry Holt, 1996. 32p. Grades K–5. F, K–4, 2–5.

Examines the many forms of wildlife that can be displaced if their environment is destroyed by development and discusses how communities and schools can provide space for them to live. Includes Bibliography; Sources.

Florian, Douglas. *In the Swim.* San Diego, Calif.: Harcourt, 1997. 48p. Grades K–6 C, K–4, 1–3.

Continuing his series of nature poetry books, Florian amusingly captures twenty-one underwater creatures in words and watercolors. Also, *Beast Feast* (animal poems); *On the Wing* (bird poems).

———. *Insectlopedia.* San Diego, Calif.: Harcourt, 1998. 48p. Grades K–6. C, K–4, 1–3.

Fabulous art and witty poems introduce twenty-one insects.

Fowler, Allen. *Rookie Read-About Science.* Series. New York: Children's Pr., 1998, 1999. 32p. Grades K–2. C, K–4, 1–3; B, K–4, 1–3.

Full-color photos and minimal text involve young readers as they discover intriguing facts about the world around them. Words You Know uses pictures to associate words. Titles are:

*Inside an Ant Colony*

*Good Mushrooms and Bad Toadstools*

*A Snail's Pace*

*Can You See the Wind?*

*Arms and Legs and Other Limbs*

Frazier, Debra. *Out of the Ocean.* San Diego, Calif.: Harcourt, 1997. 32p. Grades PreK–2. C, K–4, 1–3.

A young girl and her mother walk along the beach and marvel at the treasures cast up by the sea and wonder of the world around them. Photographs and drawings make for easy identification.

Friedhoffer, Bob. *Physics Lab in a Hardware Store.* New York: Watts, 1996. 112p. Grades 5–12. B, 5–8, 1–3; B, 9–12, 4–6.

Examines such topics in physics as mass, weight, gravity, buoyancy, and pressure with experiments using common household tools. Looks at the physics behind sandpaper, the wheelbarrow, a window shade, a pry bar, and many other tools and gadgets that can be found in a hardware store. Also, *Science Lab in a Supermarket,* 1998.

Gaffney, Timothy. *Grandpa Takes Me to the Moon.* Pictures by Barry Root. New York: Tambourine, 1996. 32p. PreK–3. D, K–4, 2.

A child whose grandfather was an astronaut always asks Grandpa for a bedtime story in which the two of them blast off for the moon together.

Galan, Mark. *There's Still Time: The Success of the Endangered Species Act.* Washington, D.C.: National Geographic Society, 1997. 40p. Grades K–6. C, 5–8, 1–5.

Full-color photos introduce animals, birds, and plants on the road to recovery. An introduction for younger children and a springboard to further research for older students. Index.

Ganeri, Anita. *The Hunt for Food. Life Cycles* series. Brookfield, Conn.: Millbrook, 1997. 30p. Grades 2–5. C, K–4, 1–3.

Describes the interdependence among plants and animals living in a meadow environment, from spring to winter.

———. *Inside the Body* (A Lift-the-Flap Book). New York: Dorling Kindersley, 1996. 10p. Grades 4–8. C, 5–8, 1.

Open the flaps to reveal the secret inner workings of the human body.

Gardner, Robert. *Where on Earth Am I?* New York: Watts, 1996. 160p. Grades 5–12. D, 5–8, 1–3.

Offers a variety of investigations, activities, and projects explaining how humans discovered Earth's position in the universe and how we can find our own location using maps, compasses, the sun, and the stars. Glossary; Further Reading; Index.

George, Kristine O'Connell. *Old Elm Speaks: Tree Poems*. Illustrated by Kate Kiesler. New York: Clarion, 1998. 48p. Grades K–6. C, K–4, 1–3; Language Arts.

A collection of short, simple poems that present images relating to trees in various circumstances and throughout the seasons.

George, Jean Craighead. *Look to the North: A Wolf Puppy Diary*. Illustrated by Lucia Washburn. New York: HarperCollins, 1997. 32p. Grades K–2. C, K–4, 1.

The life cycle of three pups from day one to ten and a half months.

Gerstenfeld, Sheldon. *Zoo Clues*. Illustrated by Eldon Doty. New York: Viking, 1991. 113p. Grades 3–6. C, K–4, 1 & 3.

Fascinating facts of interest to anyone. Illustrations and text show children what animals to look for when they go to the zoo.

Gibbons, Gail. *Gulls . . . Gulls . . . Gulls . . .* New York: Holiday, 1997. 32p. Grades K–3. C, K–4, 1–3.

Bright, uncluttered illustrations, interesting facts, and easy text describing the life cycle, behavior patterns, and habits of various species of gulls make this a good beginning reader. Contains gull facts in appendix. Also:

*The Moon Book,* 1997
*Cats,* 1996
*Dogs,* 1996
*Deserts,* 1996

———. *The Honey Makers.* New York: Morrow, 1997. unp. Grades K–2. C, K–4, 2–3.

Covers the physical structure of honeybees and how they live in colonies, as well as how they produce honey and are managed by beekeepers.

Gibson, Gary. *Science for Fun Experiments.* Brookfield, Conn.: Copper Beech Books, 1996. 224p. Grades K–8. A, K–4, 1–2; A, 5–8, 1–2; B, K–4, 1–3; B, 5–8, 1–3.

Provides instructions for a selection of hands-on science experiments introducing basic scientific principles in magnetism, electricity, and water.

Glaser, Linda. *Compost! Growing Gardens from Your Garbage.* Pictures by Anca Hariton. Brookfield, Conn.: Millbrook, 1996. 32p. Grades PreK–3. F, K–4, 3 & 4.

Through the eyes of a little girl, this book describes what composting is, what it does, and how to go about it.

Glover, David. *Simple Machines.* Series. Des Plaines, Ill.: Heinemann, 1997. 24p. Grades PreK–4. B, K–4, 1–2; B, 5–8, 1–3.

The titles in this series examine basic technology at work around us and relate the technology to common machines in our everyday world.

*Levers*
*Pulleys and Gears*
*Ramps and Wedges*
*Screws*
*Springs*
*Wheels and Cranks*

Graham, Ian. *Energy Forever?* Series. Austin, Tex.: Raintree Steck-Vaughn, 1999. 48p. Grades 4–8. B, 5–8, 1–3.

This series examines future energy needs, fuel availability, and alternative energy sources. Full-color photographs and clear diagrams make for easy understanding. Further information; Glossary; Index. Titles are:

*Solar Power*
*Water Power*
*Wind Power*
*Fossil Fuels*
*Geothermal and Bio-Energy*
*Nuclear Power*

Grazzini, Francesca. *I Want to Know.* Series. Illustrated by Chiara Carrer. Brooklyn, N.Y.: Kane/Miller, 1996. 26p. Grades PreK–K. A, K–4, 1–2.

Titles are:

*Flower, Why Do You Smell So Nice?*
*Wind, What Makes You Move?*
*Rain, Where Do You Come From?*
*Sun, Where Do You Go?*

Greenaway, Theresa. *The Really Horrible Guides.* Series. New York: Dorling Kindersley, 1996. 24p. Grades K–3. C, K–4, 1.

Each *Really Horrible Guide* is a stunning insight into the beasts children love to loathe. Filled with fascinating background information and amazing, full-color photography, each book teaches young readers all about the amazing lives of hundreds of strange and intriguing creatures. Titles are:

*The Really Fearsome Blood-Loving Vampire Bat and Other Creatures with Strange Eating Habits*
*The Really Hairy Scary Spider and Other Creatures with Lots of Legs*
*The Really Horrible Horned Toad and Other Cold, Clammy Creatures*
*The Really Wicked Droning Wasp and Other Things That Bite and Sting*

———. *Weird Creatures of the Wild.* New York: Dorling Kindersley, 1997. 21p. Grades K–3. C, K–4, 1–3.

A close-up look at some of the most extraordinary creatures of the animal kingdom: cassowary, tapir, Komodo dragon, and more.

Amazing facts and scale drawings show the size of each animal compared to a person.

Greenlaw, M. Jean. *Welcome to the Stock Show.* New York: Lodestar, 1997. 48p. Grades 3–6. C, K–4, 1 & 2.

Clear photos capture the tasks involved in preparing animals for a livestock show—goats, a calf, and rabbits. Bibliography; Glossary.

Guiberson, Brenda. *Into the Sea.* New York: Henry Holt, 1996. 32p. Grades K–3. C, K–4, 1–2.

Follows the life of a sea turtle from its hatching on a beach through its years in the sea, and its return to land to lay its eggs.

Haddon, Mark. *The Sea of Tranquility.* Illustrated by Christian Birmingham. San Diego, Calif.: Harcourt, 1996. 28p. Grades 1–3. D, K–4, 2.

A man remembers his boyhood fascination with the moon and the night astronauts first bounced through the dust in the Sea of Tranquility.

Hall, Drew. *The Surprise Garden.* Illustrated by Shari Halpern. New York: Blue Sky/Scholastic, 1998. 32p. PreSch–1. C, K–4, 1–3.

Bright collages and simple text enhance this modern version of planting a carrot seed. Appendix shows seeds, pictures, markers, etc.

Hanly, Sheila. *The Big Book of Animals.* New York: Dorling Kindersley, 1997. 48p. Grades K–4. C, K–4, 1–3.

Over 365 animals shown by habitat along with a section on domestic animals. Index.

Hansen, Mary Elizabeth. *Snug.* New York: Simon & Schuster, 1997. 32p. Grades PreK–1. C, K–4, 1–3.

A mother bear patiently rescues her mischievous cub when he gets into one difficulty after another while playing instead of learning to hunt.

Hausherr, Rosmrie. *What Food Is This?* New York: Scholastic, 1994. 40p. Grades K–4. C, K–4, 1–3.

In question-and-answer format, discusses 18 different foods representing the food groups and provides additional information on nutrition, healthy eating habits, and meal preparation with kids in mind.

Hewitt, Sally. *It's Science.* Series. New York: Children's Pr., 1998. 32p. Grades PreK–3. B, K–4, 1–3.

Introduces students to the basic concepts of physical science covered in the first years of school while encouraging awareness of scientific processes at work in the world around us. "Try It Out!," "Think About It!" and "Look Again" sections make this a must for every elementary library. Titles are:

*Solid, Liquid, or Gas?*
*Full of Energy*
*Forces around Us*
*Machines We Use*

*The Highlights Big Book of Science Secrets: Puzzles, Projects, Experiments, and Challenges Galore.* Honesdale, Pa.: Boyd Mills Pr., 1997. 144p. Grades K–8. B, K–4, 1–3; B, 5–8, 1–3.

Organized in steps with clear illustrations, this book by *Highlights for Children* presents a broad range of topics including food, weather, water, senses, light. Includes index.

Himmelman, John. *Nature Upclose.* Series. New York: Children's Pr., 1998. 32p. Grades K–2. C, K–4, 1–3.

Follow a variety of small creatures as they go about their daily lives and through their entire life cycles. Strikingly original watercolors depict each creature's world from its own unique perspective. Simple text describes the creatures' movements and activities. Italicized words in Words You Know. Scientific name included. Titles are:

*A Ladybug's Life*
*A Salamander's Life*
*A Luna Moth's Life*
*A Slug's Life*

Hirshi, Ron. *A Wildlife Watcher's First Guide.* Series. Photographs by Thomas D. Mangelesen. New York: Cobblehill, 1997. 32p. Grades K–3. C, K–4, 1–3.

*Faces in the Mountains.* Focuses on animals that live in mountainous regions of the United States, including grizzly bears, marmots, bighorn sheep, and others. Beautiful color photos.

*Faces in the Forests.* Introduces the various animals that live in the forest of the United States, including beavers, chipmunks, spotted owls, bears, deer, bobcat, eagles, and others.

Holmes, Kevin. *Bridgestone Animals.* Series. Mankato, Minn.: Bridgestone, 1998. 24p. Grades 2–5. C, K–4, 1–3.

In addition to striking color photography the appendices are especially useful: glossary, a frog game, bibliography, useful address, Internet sites, and index! This series gives the facts about what different types of animals look like, where they live, what they eat, and who eats them.

*Frogs*
*Bees*
*Butterflies*
*Snails*
*Earthworms*

Howker, Janni. *Walk with a Wolf.* Illustrated by Sarah Fox-Favies. Cambridge, Mass.: Candlewick, 1997. 32p. Grades K–2. C, K–4, 1–3.

A lyrical text and stunning art follows a female wolf as she hunts before rejoining the pack.

*Inside Guides.* Series. New York: Dorling Kindersley, 1996, 1997. 48p. Grades 3–7. C, K–4, 1; D, K–4, 3.

Children love to see the inside of things. This series features 3-D models that are broken apart, photographed, and annotated. Titles are:

*Incredible Plants*
*Human Body*
*Animal Homes*
*Incredible Earth*
*Amazing Bugs*

*Inside the Dzanga-Sangha Rain Forest.* Compiled by Francesca Lyman. New York: Workman, 1998. 128p. Grades 5 and up. C, 5–8, 1, 3–5.

An account of the American Museum of Natural History expedition to the Dzanga-Sangha Rain Forest in Central African Republic to collect specimens for an exhibit. This is the story of how many people, working together, created a rain forest inside a museum. A fascinating book.

Jackson, Ellen. *The Book of Slime.* Brookfield, Mass.: Millbrook, 1997. 32p. Grades 2–5. C, K–4, 1.

Ellen Jackson tells you everything you have ever wanted to know about things that are oily, greasy, goopy, and gross. Describes some animals and plants that are slimy. Includes recipe for edible slime and a collection of slime jokes.

Jenkins, Martin. *Chameleons are Cool.* Illustrated by Sue Schields. Cambridge, Mass.: Candlewick, 1997. 29p. Grades PreK–2. C, K–4, 1–3.

The appealing colorful art and the amusing text support the author's contention that chameleons make cool subjects for study.

Jenkins, Steve. *What Do You Do When Something Wants to Eat You?* Boston: Houghton, 1997. 32p. Grades PreK–2. C, K–4, 1.

Colorful collages introduce the defenses used by various animals.

Johnston, Tony. *An Old Shell: Poems of the Galapagos.* Pictures by Tom Pohrt. New York: Farrar, Straus and Giroux, 1999. 51p. Grades 2–6. C, K–4, 1–3.

A collection of poems exploring and celebrating the Galapagos Islands and their various animals.

Kennedy, Dorothy; ed. *Make Things Fly, Poems about the Wind.* New York: McElderry Books, 1998. 32p. Grades K–3. Language Arts.

A collection of poems describing the wind by such writers as Lilian Moore, John Ciardi, Christina Rossetti, A. A. Milne, and Eve Merriam.

Kite, Patricia. *Blood-Feeding Bugs and Beasts.* Brookfield, Mass.: Millbrook, 1996. 48p. Grades 2–4. C, K–4, 1.

Presents information about such creatures as lice, chiggers, fleas, ticks, and vampire bats. Includes Bibliography; Index.

———. *Dandelion.* Illustrated by Anca Hariton. Brookfield, Mass.: Millbrook, 1998. 32p. Grades PreK–3. C, K–4, 1–3.

Some amazing facts about one of the most common wildflowers. Colorful pictures.

Kittinger, Jo S. *Dead Log Alive! First Books—Biology.* Series. New York: Watts, 1996. 64p. Grades 5–8. A, 5–8, 1–2.

Describes the variety of animal and plant life found on, in, and around dead logs, and explains the role that dying trees play in nature's cycles. Bibliography; Index. Also by the same author:

*Lives Intertwined: Relationships between Plants and Animals*

Krupinski, Loretta. *Into the Woods: A Woodland Scrapbook.* New York: Harper, 1997. 32p. Grades K–3. C, K–4, 1–3.

This colorful book with hand printed fascinating tidbits about forest animals and plants would inspire young readers to make their own scrapbooks. A collection of facts and fiction about the woods.

Laser, Michael. *The Rain.* Illustrated by Jeffrey Green. New York: Simon & Schuster, 1997. unp. D, K–4, 1.

A lyrical introduction to a unit on weather with rich pastels recording people's reactions to the rain in the city and in the country.

Lasky, Kathryn. *The Most Beautiful Roof in the World: Exploring the Rainforest Canopy.* Photographs by Christopher Knight. San Diego, Calif.: Harcourt, 1997. unp. Grades 3–6. C, K–4, 1–3; C, 5–8, 4 & 5.

Describes the work of Meg Lowman in the rain forest canopy, an area unexplored until the last ten years and home to previously unknown species of plants and animals.

———. *Shadows in the Dawn: The Lemurs of Madagascar.* San Diego, Calif.: Harcourt, 1998. 63p. Grades 4–8. C, 5–8, 1–5.

Text and photographs follow primatologist Alison Jolly and a group of lemurs on the island of Madagascar, presenting the appearance, behavior, and social structure of this group of primates.

Lauber, Patricia. *Flood: Wrestling with the Mississippi.* Washington, D.C.: National Geographic Society, 1996. 64p. Grades 3–6. D, K–4, 1, 3.

Describes the history of flooding of the Mississippi River, focuses on the 1927 and 1993 floods, the effects on people near the river, and efforts to avoid flooding. Dramatic photos. Index.

———. *The True-or-False Book of Cats.* Illustrated by Rosalyn Schanzer. Washington, D.C.: National Geographic Society, 1998. 32p. Grades K–4. C, K–4, 1–3.

Truths behind such beliefs as cats can see in the darkness, cats can't see color, etc. Simple, with colorful illustrations.

Legg, Gerald. *Lifecycles.* Series. Illustrated by Carolyn Scrace. New York: Watts, 1997. 32p. Grades K–3. C, K–4, 1–3.

Each title traces the growth process of a particular plant or animal using simple, clear texts and stunning detailed, full-color artwork. Includes glossary, index, fact section, and timeline.

*From Tadpole to Frog* | *From Seed to Sunflower*
*From Caterpillar to Butterfly* | *From Egg to Chicken*

Lember, Barbara. *The Shell Book.* Boston: Houghton, 1997. 32p. Grades K–3. E, K–4, 1; D, K–4, 1.

A brief identification of fourteen common shells, including the lion's paw, Atlantic cockle, and Katharine's chiton.

Lesser, Carolyn. *Dig Hole, Soft Mole.* San Diego, Calif.: Harcourt, 1996. 32p. Grades PreK–2. C, K–4, 1–3.

A mole travels underground and underwater, exploring marsh and pond.

———. *The Goodnight Circle.* San Diego, Calif.: Harcourt, 1984. unp. Grades K–3. C, K–4, 1–3.

Describes the activities of a variety of animals from sunset to sunrise.

———. *Spots: Counting Creatures from Sky to Sea.* Illustrations by Laura Regan. San Diego, Calif.: Harcourt, 1999. 32p. Grades PreK–2. C, K–4, 1; Language Arts; Math.

Spotted animals from around the world, including a leopard ray, ringed seals, reticulated giraffes, and tundra butterflies, introduce the numbers from one to ten. Descriptive adjectives and gerunds.

———. *Storm on the Desert.* Illustrated by Ted Rand. San Diego, Calif.: Harcourt, 1997. 32p. Grades K–3. C, K–4, 1–3.

The flora and fauna of a southwest desert before and after a short and violent thunderstorm depicted in vivid art and lyrical text.

*Let's-Read-and-Find-Out Science.* Series. New York: HarperCollins, 1996, 1997. 32p.

Stage 1 books explain simple and easily observable science concepts. Stage 2 books explore more challenging concepts and include hands-on activities that children can do themselves. Authors and titles are:

Branley, Franklyn. *Magnets*. Illustrated by True Kelley. Stage 1. Grades K–3. B, K–4, 1–3.

Bancroft, Henrietta, and Richard Van Gelder. *Animals in Winter.* Illustrated by Helen Davie. Stage 1. Grades PreK–2.

Describes how animals cope with winter.

Pfeffer, Wendy. *From Tadpole to Frog*. Illustrated by Holly Keller. Stage 1. Grades K–3. C, K–4, 2.

Simple text and clear pictures capture the life cycle of the frog and its metamorphosis from tadpole to adult.

Lindbergh, Reeve. *North Country Spring*. Paintings by Liz Sivertson. Boston: Houghton, 1997. 32p. Grades PreK–2. C, K–4, 1; D, K–4, 3.

In rhymed text, spring awakens all nature. Includes section with facts about animal behavior.

Llewellyn, Claire. *Our Planet Earth*. New York: Scholastic, 1997. 78p. Grades 2–4. A, K–4, 1–2.

Beginning reference book with scores of color photos and two pages of interesting facts per topic. Looks at earth, how it was formed, the way it changes, and some of the plants and animals that live there.

Locker, Thomas. *Water Dance*. San Diego, Calif.: Harcourt, 1997. 32p. Grades 1–3. D, K–4, 1–3.

Stunning paintings and lyrical text introduce simple facts about the water cycle. Includes factual information on the water cycle.

Lockwood, C. C. *C. C. Lockwood's Louisiana Nature Guide*. Baton Rouge, La.: LSU Pr., 1995. 92p. All ages.

A MUST for all libraries. The fantastic color photos captivate and educate all readers.

Loewer, Peter. *Pond Water Zoo*. Illustrated by Jean Jenkins. New York: Atheneum, 1996. 89p. Grades 3–8. C, K–4, 1–3; C, 5–8, 4 & 5.

Examines the many different microscopic organisms that exist in ponds and describes their function and role in nature.

London, Jonathan. *Baby Whale's Journey*. Illustrated by Jon Van Zyle. San Francisco: Chronicle, 1999. 40p. Grades PreK–2. C, K–4, 1–3.

Off the coast of Mexico a whale is born. Baby Whale follows her mother like a shadow and, surrounded by a protective pod, begins to learn and grow. As moons come and go, she learns the way of whales and the sea. An afterword and reader's guide provide background, activities, and discussion guide.

———. *Ice Bear and Little Fox.* Paintings by Daniel San Souci. New York: Dutton, 1998. 40p. Grades K–2. C, K–4, 1–3.

Describes how a polar bear and the little fox that follows it survive over the course of a year in the Arctic. Includes afterword with facts about arctic animals and Inuit people.

Lovett, Sarah. *Extremely Weird.* Series. Santa Fe, N.M.: John Muir Publications. 32p. Grades 1–3. C, K–4, 1–3.

Includes a detailed and entertaining description of the organism along with full-color photographs. Intended to instill a greater awareness of and respect for the wonderful variety of life on our planet. Glossarized index includes scientific index. Titles are:

*Bats,* 1991

*Sea Creatures,* 1997

*Animal Hunters,* 1992

Lyon, George Ella. *Counting on the Woods.* Photos by Ann Olson. New York: Dorling Kindersley, 1998. 32p. Grades PreK–2. C, K–4, 1–3.

Uses rhyme to enumerate and describe natural objects seen while walking through the woods. Illustrated with exceptional photography.

Maass, Robert. *Garden.* New York: Henry Holt, 1998. 32p. Grades PreK–2. C, K–4, 1–3.

Exquisite color photos and lyrical text reproduce the beauty and harmony of different gardens. A pleasing introduction for young readers and listeners. Italicized words in glossary.

Madgwick, Wendy. *Science Starters.* Series. Austin, Tex.: Raintree Steck-Vaughn, 1999. 32p. Grades K–4. B, K–4, 1–3; C, K–4, 1–3; D, K–4, 1.

A bright and colorful series for the beginning scientist. Each title contains a dozen or more activities and the step-by-step instructions and ingredients for parents and teachers. These books make science learning fun. Glossary; Further Reading; Index. Titles are:

| | |
|---|---|
| *Light and Dark* | *Magnets and Sparks* |
| *Living Things* | *On the Move* |
| *Super Materials* | *Up in the Air* |
| *Super Sound* | *Water Play* |

*Make It Work! Science.* Series. Chicago: World Book, 1996. 48p. Grades 2–5. All standards.

This hands-on approach to science series is ideal for the lower elementary students. Imaginative activities and experiments teach children basic science process skills. Glossary; Index. Titles are:

| | |
|---|---|
| *Body* | *Machines* |
| *Building* | *Photography* |
| *Dinosaurs* | *Plants* |
| *Earth* | *Ships* |
| *Electricity* | *Sound* |
| *Flight* | *Time* |
| *Insects* | *Universe* |

Markle, Sandra and William. *Gone Forever! An Alphabet of Extinct Animals.* Illustrated by Felipe Davalos. New York: Atheneum, 1997. 40p. Grades 2–5. C, K–4, 1–3.

Describes, for each letter of the alphabet, an animal that lived much more recently than dinosaurs and that is now extinct. Graphic art.

Marshall, Elizabeth. *The Human Genome Project.* New York: Watts, 1996. 128p. Grades 5–10. C, 5–8, 1–2; G, 5–8, 1.

Understanding the human genome could lead to treatments for many life-threatening diseases. The fifteen year, multimillion dollar effort involves researchers around the world. Glossary; Source Notes; Bibliography; Index.

Martin, Jacqueline. *The Green Truck Garden Giveaway!* Illustrated by Alec Gillman. New York: Simon & Schuster, 1997. 32p. Grades K–5. C, K–4, 1–3.

When two people pass out seeds and gardening supplies, neighbors who claim to have no interest in gardening are transformed into a community of gardeners. Includes helpful information on gardening. This story is so satisfying with all kinds of intriguing sidebars.

Mason, Cherie. *Wild Fox.* Camden, Maine: Down East, 1993. 32p. All grades. C, K–4, 1–3.

A true nature story of kindness, sensitivity, and respect gives this beautifully written and illustrated book an appeal for all ages. Absolutely enchanting.

Maynard, Caitlin and Thane. *Rain Forest and Reefs: A Kid's Eye View of the Tropics.* Photographs by Stan Rullman. New York: Watts, 1996. 64p. Grades 5–8. C, 5–8, 4 & 5.

Describes a field trip that eighteen students and seven adults took to study the ecosystems of a tropical rain forest and a coral reef in Belize. Glossary; Index.

Maze, Stephanie. *I Want to Be . . .* Series. San Diego, Calif.: Harcourt, 1997. 48p. Grades 4–8.

Provides an overview of what is involved in preparing for and choosing various scientific careers.

*A Veterinarian* *An Astronaut*

McKenna, Virginia. *Back to the Blue.* Illustrated by Ian Andres. Brookfield, Conn.: Millbrook, 1997. 40p. Grades 1–5. C, K–4, 1–3.

Fiction and fact are combined in this story of a captive dolphins being returned to the sea. Includes a section with facts and photographs about the real rescue effort on which the story was based.

McMillan, Bruce. *Wild Flamingo.* Boston: Houghton, 1997. 32p. Grades 2–6. C, K–4, 1–3.

Fascinating photos and intriguing facts about the world's largest flamingos that lives on the Caribbean island of Boncure. Bibliography; Index. About the Great Flamingo.

Meyers, Jack. *Highlights Book of Science Questions That Children Ask.* Honesdale, Pa.: Boyds Mill Pr., 1995. 255p. Grades 1–5. All standards.

Answers 350 real questions from real kids. Index.

Micucci, Charles. *The Life and Times of the Peanut.* Boston: Houghton, 1997. 32p. Grades 2–6. C, K–4, 1–3.

Peanut butter is 3000 years old! Everything you want to know about peanuts with colorful, clear illustrations.

Miller, Debbie. *A Polar Bear Journey.* Boston: Little Brown, 1997. 32p. Grades 1–4. C, K–4, 1–3.

Beautiful art chronicles the life cycle of a mother polar bear and her two cubs, from birth to their learning of survival lessons on the coast of Alaska and northwest Canada. Bear facts in appendix, with map, and scientific name.

Miller, Joe. *If the Earth . . . Were a Few Feet in Diameter.* Artwork by Wilson McLean. Shelton, Conn.: Greenwich Workshop Pr., 1998. 32p. Grades K–4. All standards; Language Arts.

The wonders of the earth are portrayed in paintings, poetic text, and a presentation of facts about planet Earth. An excellent read aloud and discussion book.

Morgan, Roland. *In the Next Three Seconds.* New York: Lodestar, 1997. 32p. All grades. A, K–4, 5–8, 9–12, 1–2.

*In the Next Three Seconds* . . . "ninety-three trees will be cut down to make the liners for disposable diapers." Hundreds of entertaining and humorous, disturbing, and astonishing predictions with explanations for making your own calculations. "In the next three hours seventeen species of life will disappear from Earth!" Predicts events that will occur in the near and distant future.

Morrison, Gordon. *Bald Eagle.* Boston: Houghton, 1998. 30p. Grades 2–6. C, K–4, 1–3; C, 5–8, 1–5.

Detailed color illustrations and readable text follow the developmental stages of two baby eagles as they hatch and grow plumage, learn to fly, and struggle to fish and hunt on their own. Accompanying annotated sketches give more in-depth information for life science and behavior.

Morrison, Taylor. *Cheetah.* New York: Henry Holt, 1998. 32p. Grades PreK–2. C, K–4, 1–3.

Describes a day in the life of a cheetah family in the Serengeti National Park.

Moser, Madeline. *Ever Heard of an Aardwolf?* San Diego, Calif.: Harcourt, 1996. 44p. Grades K–6. C, K–4, 1 & 3.

Here's the lowdown on strange, interesting, adorable, repulsive, and peculiar animals that you may never have heard of. This book presents information about twenty animals with unusual names and habitats. Glossary includes complete information on these real animals.

Moss, Jeff. *Bone Poems*. New York: Workman, 1997. 96p. Grades 3–7. Language Arts.

A joyous collection of poems for kids who love dinosaurs, plays on words, and the delightful incongruities of science and nature.

National Wildlife Federation. *The Unhuggables*. Vienna, Va.: National Wildlife Federation, 1988. All grades. C, K–4, 1–3.

Describes the physical characteristics, habitats, and natural environment of a variety of mammals, insects, and other animals people often fear, dislike, or simply ignore.

Older, Jules. *Cow*. Illustrated by Lyn Severance. Watertown, Mass.: Charlesbridge, 1977. unp. Grades PreK–2. C, K–4, 1.

A light-hearted, informative look at cows: different breeds, what they eat, how they make milk, and an assortment of other facts. Quiz at conclusion of book.

Oppenheim, Joanne. *Have You Seen Bugs?* Illustrated by Ron Broda. New York: Scholastic, 1996. 32p. Grades Presch–3. C, K–4, 1–3.

Describes in verse a variety of bugs and how they look, behave, and improve our lives. About Bugs appendix and common names of bugs listed by page number. Extraordinary art.

Otten, Charlotte. *January Rides the Wind: A Book of Months*. New York: Lothrop, 1997. 32p. Grades Presch–3. Language arts.

Unrhymed verses and big, luminous paintings of nature and of children in nature describe the twelve months.

Oxlade, Chris. *Step-by-Step Science*. Series. New York: Children's Pr., 1998. 32p. Grades 2–5.

Written in a direct and easy style, the titles in this series introduce a wide range of scientific subjects. Each topic is explained with examples from everyday life, and through the use of activities and experiments. Photographs and illustrations complement the text. Glossary; Index. Titles are:

*Energy and Movement.* B, 5–8, 1–3

*Flowering Plants.* C, K–4, 1–3

*Light and Dark.* B, K–4, 3

*Space.* D, K–4, 1–3

Parker, Steve. *The Beginner's Guide to Animal Autopsy.* Brookfield, Conn.: Copper Beech Books, 1997. 48p. Grades 5–8. C, 5–8, 1–3.

Investigates the internal workings of animals, with drawings of dissections and cartoons illustrating how the animals' bodies work. Includes Index; Glossary; Animal Classification.

———. *Professor Protein's Fitness, Health, Hygiene, and Relaxation Tonic.* Brookfield, Conn.: Copper Beech Books, 1996. 48p. Grades 5+ F, K–4, 1.

Professor Protein, Carol Calorie, and their friends tell how to get fit, stay healthy, keep clean, relax and rest, and enjoy life through the best foods, drink, exercises, and sports.

Pascoe, Elaine. *Nature Close-Up.* Series. Photographs by Dwight Kuhn. Woodbridge, Conn.: Blackbirch Pr., 1997. 48p. Grades 3–6. C, K–4, 1–3; C, 5–8, 1–3.

Intimate view of nature through photographs. Text brings the facts to life through hands-on investigation and observation. Bibliography; Index. Titles are:

*Earthworms*

*Butterflies and Moths*

*Seeds and Seedlings*

*Tadpoles*

Patent, Dorothy Hinshaw. *Biodiversity.* Photographs by William Munoz. New York: Clarion, 1996. 109p. Grades 5–10. A, 5–8, 1–2; C, 5–8, 3–5; C, 9–12, 4–6.

Provides a global perspective on environmental issues while demonstrating the concept that encompasses the many forms of life on earth and their interdependence on one another for survival.

———. *Prairies.* Photographs by William Munoz. New York: Holiday, 1996. 40p. Grades 3–6. C, 5–8, 1; F, 5–8, 2.

Color photographs reflect the ecology and conservation of the North American prairie. Index.

———. *Back to the Wild.* Photographs by William Munoz. San Diego, Calif.: Harcourt, 1997. 69p. Grades 3–8. C, 5–8, 1–5.

An award-winning duo's latest collaboration examines the breeding of endangered species and their introduction into the wild. Describes efforts to save endangered animals from extinction by breeding them in captivity, teaching them survival skills, and releasing them into the wild.

Perry, Phyllis. *Armor to Venom: Animal Defenses. First Books* Series. New York: Watts, 1997. 64p. Grades 4–6. C, 5–8, 1, 5.

Describes how animals survive by using their armor, camouflage, horns, stings, and other natural protective devices and approaches. Boldface type indicates word is in glossary.

Peterson, Cris. *Harvest Year.* Photographs by Alvis Upitis. Honesdale, Pa.: Boyd Mills Pr., 1996. 32p. Grades K–3. A, K–4, 1–2; C, K–4, 3.

A photographic essay about foods that are harvested year-round in the United States.

Pomeroy, Diana. *Wildflower ABC: An Alphabet of Potato Prints.* San Diego, Calif.: Harcourt, 1997. 32p. Grades K–3. C, K–4, 1.

Upper and lower case letters, presents a glossary of information for each flower.

Porte, Barbara Ann. *Tale of a Tadpole.* New York: Orchard, 1997. 28p. Grades PreK–2. C, K–4, 1–3.

An enchanting story of a young girl who watches her pet tadpole turn into a toad.

Pringle, Lawrence. *An Extraordinary Life: The Story of a Monarch Butterfly.* Painting by Bob Marshall. New York: Orchard, 1997. 64p. Grades 4–9. C, 5–8, 1–3.

Introduces the life cycle, feeding habits, migration, predators, and mating of the monarch butterfly through the observations of one articular monarch named Danaus. Stunning art with informative sidebars and descriptive captions providing a wealth of interconnections with other animals, geography, people, climate, and history. Further Reading; Index; How to Raise a Monarch Butterfly; Maps.

Raines, Kenneth and Bruce Rusell. *Guide to Microlife.* New York: Watts, 1996. 287p. Grades 5–10. C, 5–8, 1–5.

Serves as a guide for the identification of microorganisms and provides information about microlife and how they affect other lifeforms, including humans. Color photographs, illustrations, and detailed descriptions. Bibliography; Index of Organisms; General Index; Appendices.

Riley, Peter. *Straightforward Science.* Series. New York: Watts, 1998, 1999. 32p. Grades K–4.

A straightforward introduction to science that explains the main scientific principles and shows how they work from our everyday world to outer space. Clear photographs and diagrams complement the text. Glossary; Index. Titles are:

*Electricity.* B, K–4, 3

*Materials and Processes.* B, K–4, 1–3

*Forces and Movement.* B, K–4, 1–3

*Food Chains.* C, K–4, 1, 3

*Magnetism.* B, K–4, 3

*Light and Color.* B, K–4, 3

*The Earth in Space.* D, K–4, 1–3

*Plant Life.* C, K–4, 1–3

Ring, Elizabeth. *What Rot!: Nature's Mighty Recycler.* Brookfield, Conn.: Millbrook, 1996. 32p. Grades K–3. C, K–4, 1 & 2.

Text and photos show how rot and all the tiny organisms that cause it to maintain the cycle of life. Glossary.

Robson, Pam. *What's For Lunch?* Series. New York: Children's Pr., 1997. 32p. Grades K–2. C, K–4, 1–2.

Using clear text and striking photographs, the series provides a look at foods around the world—how they are grown, made, and eaten.

A glossary explains unfamiliar words. Titles are:

*Corn*
*Rice*

Rosen, Michael, et al. *Down to Earth: Garden Secrets! Garden Stories! Garden Projects You Can Do!* San Diego, Calif.: Harcourt, 1998. 64p. All grades. C, K–4, 1–3; C, 5–8, 1–5.

Rosen and forty-one noted children's authors and illustrators share their gardening experiences. Includes various activities and recipes relating to gardening. Oversized format with beautiful illustrations.

Royston, Angela. *Life Cycle of a . . .* Series. Des Plaines, Ill.: Heinemann, 1998. 32p. Grades K–3. C, K–4, 1–3.

This series takes an in-depth look at the lives of some familiar plants and animals. Each book contains colorful photographs, illustrated timeline of the stages of development, and fact files. Glossary; Index; Bibliography.

*Apple*
*Bean*
*Butterfly*
*Chicken*
*Frog*
*Guinea Pig*
*Kangaroo*
*Sunflower*

Rubel, David. *Scholastic Kids' Encyclopedia* series. New York: Scholastic, 1995. 192p. Grades 3–6. A, 5–8, 1–2.

Hundreds of visuals and facts about five topics: astronomy, biology, earth science, human body, physics and chemistry. A must reference book for elementary and middle school libraries. Index.

Rucki, Ani. *When the Earth Wakes.* New York: Scholastic, 1998. 32p. Grades PreSch–K. D, K–4, 1.

Lovely art introduces the seasons and how nature reacts to each.

Ryan, Pam. *A Pinky Is a Baby Mouse, and Other Baby Animal Names.* Illustrated by Diane DeGroat. New York: Hyperion, 1997. 32p. PreK–2. C, K–4, 1 & 3.

Rhyming text and colorful pictures of animal babies explains the different names by which various baby animals are known. Glossary lists 100 animals and their baby names, e.g., cockroach/nymph.

Ryden, Hope. *ABC of Crawlers and Flyers.* New York: Clarion, 1996. 31p. Grades K–2. C, K–4, 1.

Spectacular color photographs along with text present a different insect for each letter of the alphabet, from ants and aphids to zebra longwing.

Ryder, Joanne. *Shark in the Sea.* Illustrated by Michael Rothman. New York: Morrow, 1997. Grades K–2. C, K–4, 3.

One fall day a great white shark, hungry and alone, glides unseen, watching and waiting for someone unwary to become his meal.

*Sam's Science.* Series. Cambridge, Mass.: Candlewick, 1998, 1999. 32p. Grades PreK–3. C, K–4, 1–3.

The titles in this fun series explore topics that kids find fascinating. Bold, bright illustrations and lively, conversational text provide a clear and amusing presentation of facts that kids can really enjoy. Titles are:

*Sam's Science: I Know How We Fight Germs*

*Sam's Science: I Know Where My Food Goes*

Sandburg, Carl. *Grassroots.* Paintings by Wendell Minor. San Diego, Calif.: Harcourt, 1998. 32p. Grades 3–8. Language Arts.

The nature of the Midwest expressed in poetry and art—a stunning combination.

Sandeman, Anna. *Body Books.* Series. Brookfield, Conn.: Copper Beach Books, 1996. 32p. Grades K–3. C, K–4, 1.

Describes the functions of the various systems of the body with a focus on learning and remembering as well as problems that occur within the system. Meticulously researched by medical doctors and classroom teachers, this series offers young students a lively introduction to how the body works. Did You Know sections provide amazing facts. Glossary; Index. Titles are:

*Brain*

*Eating*

*Bones*

*Breathing*

*Senses*

*Blood*

*Babies*

*Skin, Teeth, and Hair*

Saunders-Smith, Gail. *Weather.* Series. Mankato, Minn.: Pebble, 1998. 24 p. Grades K–3. D, K–4, 3.

For the beginning reader a science series that introduces interesting facts with a glossary, bibliography, and Internet sites. Illustrated with clear color photographs. Simple text introduce students to expository language while answering students' weather questions. Titles are:

*Clouds*
*Lightning*
*Rain*
*Sunshine*

———. *Growing Flowers.* Series. Mankato, Minn.: Pebble, 1998, 24p. Grades K–3. C, K–4, 1–3.

In the same format and equally effective as the previous series: *Weather.* In this series, early readers learn about the natural world by studying a variety of flowers. Titles are:

*Flowers*
*Leaves*
*Seeds*
*Stems*

Schlein, Miriam. *What's a Penguin Doing in a Place like This?* Brookfield, Conn.: Millbrook, 1997. 48p. Grades 2–4. C, K–4, 1–3.

Outlines the varied worldwide habitats, differences, and common traits of seventeen kinds of penguins.

*Science All around Me.* Series. Des Plaines, Ill.: Heinemann, 1997. 24p. Grades K–3. B, K–4, 1–3.

Basic science concepts are presented with vivid photographs, clear text, and a variety of simple-to-do experiments. Glossary; Index; Further Readings. Titles are:

*Electricity*
*Magnets*
*Floating and Sinking*
*Matter*
*Forces*
*Moving*
*Hot and Cold*
*Sound and Light*

*Science Explorer: The Best Family Activities from the World's Favorite Hands-On Science Museum.* An Exploratorium-At-Home Book. New York: Henry Holt, 1996. 128p. Grades K–8. C, K–4, 1–2; C, 5–8, 1–2; E, K–4, 1–3.

The first ever family-oriented collection from San Francisco's Exploratorium Museum.

*The Science Explorer Out and About: Fantastic Science Experiments Your Family Can Do Anywhere.* New York: Henry Holt, 1997. 128p. Grades 4–8. A, 5–8, 1–2; B, 5–8, 1–3; E, 5–8, 1–3.

San Francisco's world-renowned Exploratorium knows how to bring parents and children together with smart, hands-on play. Each activity is accompanied by an informative What's Going On? that explains the underlying scientific concept and gives tips on how to create your own related experiments.

*Science of the Past.* Series. New York: Watts, 1998, 1999. 64p. Grades 4–8. All standards; social studies standards.

Each title in the series explains it all. Glossary; Resources (Books and Internet sites); Index. Titles are:

*Science in Ancient China*
*Science in Ancient Egypt*
*Science in Ancient Greece*
*Science in Ancient Rome*
*Science in Early Islamic Culture*
*Science in Mesopotamia*

*Science Topics.* Series. Des Plaines, Ill.: Heinemann, 2000. 32p. Grades 3–7. B, 5–8, 1–3; C, 5–8, 1–5; D, 5–8, 1–3.

This series provides an in depth look at major science concepts. In addition to the facts, common questions are answered. Common happenings and events ground the facts to everyday experiences. Includes index, glossary, and more books to read. Titles are:

*The Earth and Beyond*
*Energy*
*The Human Body*
*The Living World*
*Matter*
*Forces and Motion*
*Electricity and Magnetism*
*Light and Sound*
*Ecosystems & Environment*
*Chemicals in Action*

*Secrets of Space.* Series. Brookfield, Conn.: Twenty-First Century Books, 1996, 1997. 64p. Grades 5–8. D, 5–8, 1–3; E, 5–8, 1–2.

Explores both commonly studies space subjects—such as the solar system and comets—as well as more "offbeat" ones—such as UFO's and ETs. Titles are:

*Asteroids, Comets, and Meteors*

*The Sun and the Solar System*

*Unidentified Flying Objects and Extraterrestrial Life*

Seibert, Patricia. *Toad Overload: A True Tale of Nature Knocked Off Balance in Australia.* Illustrated by Jan Davey Ellis. Brookfield, Conn.: Millbrook, 1996. 32p. Grades K–3. C, K–4, 3.

Explains what happened when giant toads were brought to Australia to help control beetles that ate the sugar cane crop. Includes information on the physical characteristics and habits of this species of toads.

Sierra, Judy. *Antarctic Antics: A Book of Penguin Poems.* Illustrated by Jose Aruego and Adriane Dewey. San Diego, Calif.: Harcourt, 1998. 32p. Grades K–3. C, K–4, 1–3.

These amusing poems are actually based on the real lives and habits of emperor penguins; comical, colorful art.

Silverstein, Alvin, Virginia Silverstein and, Laura Silverstein Nunn. *My Health.* Series. New York: Watts, 1999. 45p. Grades 3–6. F, K–4, 1; F, 5–8, 1.

Explains how people catch colds, deal with allergies, how the body takes care of wounds, and how to take care of teeth and prevent cavities. Includes bibliography; Index. Titles are:

*Common Colds*

*Tooth Decay and Cavities*

*Allergies*

*Cuts, Scrapes, Scabs, and Scars*

Simon, Seymour. *The Brain: Our Nervous System.* New York: Morrow, 1997. 32p. Grades 3–7. C, 5–8, 1.

The various parts of the brain and the nervous system are described in interesting readable style with colorful illustrations by the author of over 150 outstanding science books for young readers.

———. *Ride The Wind: Airborne Journeys of Animals and Plants.* San Diego, Calif.: Harcourt, 1997. 40p. Grades 2–5. C, 5–8, 1–3.

An outstanding scientist tracks the migratory travels of animals and plants. Beautiful watercolors illustrates a lyrical text.

———. *Einstein Anderson, Science Detective.* Series. New York: Morrow, 1997. 92p. Grades 3–6. A, 5–8, 1–2.

Fantastic introduction to science through ten short mysteries based on scientific principles. Amusing, simple text for reluctant readers. Part of the Can you solve the mystery? series. Titles include:

*The Halloween Horror* *The Time Machine*

———. *Strange Mysteries from around the World.* New York: Morrow, 1997. 58p. Grades 3–8. A, K–4, 1–2.

Describes nine strange natural phenomena and possible explanations for them, including the day it rained frogs, an atomic explosion that occurred forty years before the atomic bomb, and an eerie crystal skull.

———. *Wild Babies.* New York: HarperCollins, 1997. Grades K–3. C, K–4, 1.

Describes the various parenting techniques of different kinds of wild animals and provides a close look at the behavior and characteristics of their offspring.

Sirett, Dawn. *The Really Amazing Animal Book.* New York: Dorling Kindersley, 1996. 24p. Grades PreK–2. C, K–4, 1.

Based on the TV series *Amazing Animals.* Colorful photos, interesting facts.

Smith, Roland and Michael J. Schmidt. *In the Forest with the Elephants.* San Diego, Calif.: Harcourt, 1998. 52p. Grades 2–6. C, K–4, 1–3.

The intriguing story of the close working relationship of men and their elephants in the jungles of Burma is told with extraordinary color photos and very interesting text.

*The Snake Book.* New York: Dorling Kindersley, 1997. unp. Grades K–5. C, K–4, 1–3; C, 5–8, 5.

Close-up photographs and vivid, flowing descriptions of these intriguing animals. Includes snake statistics.

Snedden, Robert. *YUCK! A Big Book of Little Horrors.* New York: Simon & Schuster, 1996. unp. Grades 2–8. G, K–4, 1.

Big, color photos of objects magnified from 100 times normal size to as much as 35,000 times reveal the horrors that live with us in our homes. Guaranteed to shock.

Solheim, James. *It's Disgusting and We Ate It! True Facts from around the World and throughout History.* Illustrated by Eric Brace. New York: Simon & Schuster, 1998. 37p. Grades 2–6. Language Arts.

Ugh! Guaranteed to shock and delight young readers. A collection of poems, facts, statistics, and stories about unusual foods and eating habits both contemporary and historical. Bibliography; Index of foods.

Swinburne, Stephen. *Water for One, Water for Everyone: A Counting Book of African Animals.* Brookfield, Conn.: Millbrook, 1998. 32p. Grades PreK–1. Language Arts.

A counting tale in which native animals, from one tortoise to ten elephants, arrive at a Kenyan waterhole. Glossary; Swahili terms for ten animals and ten numbers.

Toksvig, Sandi. *If I Didn't Have Elbows.* Illustrated by David Melling. New York: De Agostini Childrens Books, distributed by Stewart, Tabori and Chang, 1996. 32p. Grades K–3. C, K–4, 1.

Explains in comic-book format how the body works and postulates what might happen if certain parts were missing.

Travers, Will. *The Elephant Trunk.* Illustrated by Lawrie Taylor. Brookfield, Conn.: Millbrook, 1997. 40p. Grades 2–5. C, K–4, 1–3.

An intriguing blend of fiction and fact tells the true story of saving some elephants in Kenya who were in danger because of their destruction of cultivated crops. Includes a section with facts and photographs about the real rescue effort on which the story was based.

*Very First Things to Know About . . .* Series. New York: Workman, 1997. 32p. Grades PreK–3. C, K–4, 1–3.

An American Museum of Natural History children's book series. Each book focuses on the lives, varieties, habits, and habitat of a single animal. Detailed, realistic, full-color illustrations. Titles are:

*Ants*
*Monkeys*
*Bears*
*Frogs*

Ward, Helen. *King of the Birds.* Brookfield, Conn.: Millbrook, 1997. 36p. Grades PreK–3. C, K–4, 1.

When chaos reigns among the bird, the oldest and wisest birds declare a contest to determine who will be their king. Detailed illustrations. Common names of the birds included.

*Why Books.* Series. New York: Dorling Kindersley, 1996, 1997. 24p. Grades K–3. A, K–4, 1–2.

Why do children question at inappropriate moments? Because they can. Titles on time, food, water, earth, and more cover all the bases, leading young minds to a deeper understanding of the key concepts of their world. Titles are:

*Why Do We Laugh?* (human body)

*Why Are Zebras Black and White?* (color)

*Why Do Sunflowers Face the Sun?* (nature)

*Why Does Lightning Strike?* (weather)

*Why Do Seasons Change?* (time and seasons)

*Why Are Pineapples Prickly?* (food)

*Why Are There Waves?* (water)

*Why Do Volcanoes Erupt?* (earth)

Wick, Walter. *A Drop of Water, A Book of Science and Wonder.* New York: Scholastic, 1997. 40p. Grades K–5. B, K–4, 1; B, 5–8, 1.

Describes the characteristics of water. The camera stops the action and magnifies it so that all the amazing states of water can be observed. Experiments in appendix.

Wilkes, Angela. *My First Science Book.* New York: Dorling Kindersley, 1990. 48p. Grades PreK–2. A, B, C, K–4.

An introduction to the world of science, featuring simple experiments.

Wollard, Kathy. *How Come?* New York: Workman, 1993. 320p. Grades 5–8. A, 5–8, 1–2.

A lively omnium-gatherum of explanations to the most frequently asked questions about our world, from Why Do Stars Twinkle? to What Are Hiccups? Great for adults to share with younger children.

*Wonderwise.* Series. New York: Watts, 1997. 32p. Grades K–2. A, K–4, 1–2; C, K–4, 1, 3; E, K–4, 1–3.

This series presents an intriguing and original approach to basic science topics and concepts. Familiar images and subjects are used as starting points for extending knowledge. Next, the books make links between the familiar and the unknown, encouraging them to think and question. Titles are:

*It Takes Two*
*My Body, Your Body*
*Splish, Splash, Splosh!*
*What's under the Bed?*
*What's Up?*
*Yum-Yum!*

Woods, Shirley. *Black Nell: The Adventures of a Coyote.* Emeryville, Calif.: Groundwood, 1998. 90p. Grades 4–8. C, 5–8, 1–3.

Tells the story of Nell the coyote, from the moment of her birth through her first year of life. Action packed, scientifically accurate.

*World Explained.* Series. New York: Henry Holt, 1997. 69p. Grades 4–8. Standard Varies.

By providing clear examples and helpful explanations to untangle often complex ideas, the books in this series make their subjects accessible and truly fascinating. Intriguing information in colorful pictures and brief, clear text. Glossary; Index. Titles are:

*Inventions Explained*
*Weather Explained*
*Space Explained*
*Earth Explained*

*Worldwise.* Series. New York: Watts, 1998. 40p. Grades K–3.

This useful series is colorfully illustrated, simply written, with information about each topic conveniently arranged. Glossary; Index. Titles are:

*Volcanoes.* D, 5–8, 1
*Sharks.* C, K–4, 1–3

Wu, Norbert. *A City under the Sea.* New York: Atheneum, 1996. 28p. Grades K–5. C, K–4, 1–3; C, 5–8, 4 & 5.

Clear, colorful photos capture the life in a coral reef. Follows a sea turtle on its way back to the beach where it was born. Shows how the reef was formed and provides information on the many undersea creatures that live in and around its coral walls.

Yates, Irene. *From Birth to Death. Life Cycles.* Series. Brookfield, Conn.: Millbrook, 1997. 30p. Grades 2–5. C, K–4, 2.

Follows a year in the life of the animals and plants occupying a pond environment, showing the life cycles and interactions of the various species. Boldface text indicates word is in glossary. Glossary; Index.

Yolen, Jane. *The Originals: Animals That Time Forgot.* New York: Philomel, 1998. 32p. Grades 1–4. C, K–4, 1–3.

Thirteen poems about unusual animals that have remained almost unchanged for thousands of years although their barnyard relatives have evolved.

*Young Scientist.* Chicago: World Book, 1997. 10 volumes. Grades 2–8. All Standards.

This ten-volume set makes the young scientist in a child feel like a real scientist. The extensive use of color illustrations is an asset for the visual learner. The set features more than a 150 hands-on experiments. Each volume addresses a specific subject. Volume 10 is a study guide and a comprehensive index. A 20-volume Spanish version is also available (0-7166-6352-X).

## NOTES

1. *National Science Education Standards* (Washington, D.C.: National Academy Press, 1996).
2. Christine Y. O'Sullivan, Clyde M. Reese, and John Mazzeo, *NAEP 1996 Science Report Card for the Nation and the States* (Washington, D.C.: National Center for Education Statistics, 1997).

# 4

# Selecting Excellent Science Media

TRACY GATH

How do you select excellent science videos and software for your library? How do the materials fit in with new science standards and science education reform? What role should videos and software play in learning?

Although nothing replaces actual experience, videos can open a world not otherwise available to students. Teaching a unit on ocean life to fifth graders in Iowa? Obviously, a day trip to the ocean is out of the question, so instead a teacher can show a video, such as *Mountain in the Sea and Filming Secrets,* to take students on a virtual journey to Cocos Island off the coast of Costa Rica. Video also gives students the opportunity to meet actual scientists at work. For example, *What Is Earth Science?* shows scientists at work, and selected interviews reveal the challenges and satisfactions of their work. Each scientist is articulate and persuasive, and their presentations should help correct naive views on the part of students.

Or what about the periodic table? Students who may grow bored learning the properties of elements from a chart may be fascinated by *The Chemistry Set,* a CD-ROM that allows students to explore the periodic table in appealing ways. They can review the different states of all the elements at various temperatures, see when all the elements were discovered, view a full-screen photograph of a sample of the element, and click on the "moleculizer" icon to view three-dimensional atomic and molecular structures. Software programs can bring a subject to life and reach students who learn best through means other than simply reading—through interacting with the program, viewing pictures, hearing sounds, and manipulating data.

But how do you choose which programs are best? The criteria you use for selecting video and software programs differ from those for choosing books. Not only do you need to consider the accuracy and organization of the content, you also have to ask if the images and narration are clear. Is the sound quality acceptable? What interesting optical techniques are used to grab a viewer's attention? Is the story focused? Do the images overpower the message? With software, you need to consider these questions as well as the ease of navigation, the use of memory, compatibility with your computer, and how easily the program loads.

*Science Books and Films (SB&F),* a review journal published by the American Association for the Advancement of Science (AAAS), has been reviewing science books since 1965, films since 1975, and software since the early 1980s. *SB&F* reviewers are professionals in their fields and often point out errors that most laypeople would overlook. Reviewers are asked to evaluate materials based on accuracy; clarity of purpose; organization; scope; presentation of the processes of science; quality of photography, videography, and animation; optical techniques used; and the value of supplements.

The annotated bibliography at the end of this chapter lists recent video and software titles that were recommended by *SB&F* reviewers. For this chapter, the reviews have been categorized according to the content standards developed by the National Research Council. Although *SB&F* reviewers did not use the standards to evaluate the material, most of the reviewers are up-to-date on current standards and consider them when writing their critiques.

Other sources have evaluated materials based directly on the standards. For example, see *Resources for Teaching Middle School Science* and *Resources for Teaching Elementary School Science,* both developed by the National Science Resources Center and published by the National Academy Press. AAAS's Project 2061 has also created materials on science education reform. You can find their recommendations for excellent science materials in *Resources for Science Literacy: Professional Development.* This is Project 2061's first CD-ROM and the first professional development tool in science to focus on standards-based teaching and learning.

When evaluating videos and software, *SB&F* reviewers answer the following questions and use the answers to guide their critique.

1. Is the information accurate?
2. Is the purpose clear?
3. Is the material well organized?
4. Are the processes of science clearly and accurately presented?
5. Are conclusions, if any, valid?

Scientific accuracy of materials can be difficult for laypeople to ascertain, especially when new research makes what was once common knowledge obsolete. *SB&F* as well as other technical review publications use reviewers who are experts in the field to help determine accuracy. They can determine better than most which theories and facts are well tested, which are yet to be tested, and which are blatantly false.

Clarity of purpose and organization are other important factors to consider. Does the video or software cover too many topics too superficially? Although the program may be fun to watch or navigate, it won't meet educational objectives if it lacks focus. What message or educational objective is the producer trying to communicate? Do the images overpower the message? Does the program have a logical flow, or does it skip from one item to the next?

Science, like most human endeavors, is constantly evolving. Science standards and educational reform movements, such as AAAS's Project 2061, focus heavily on the processes of science. Science is not just a static collection of facts to be memorized, but rather a way of viewing and learning about the world around us. Videos and software should show students the excitement of research, direct their own inquiry, and expose them to new perspectives. For example, one program listed in the bibliography actually guides students in their own research: *The Geology Station* guides students to work in teams gathering, analyzing, and interpreting geological data on an undeveloped parcel of land known as Terragon Territory. The kit consists of a video; a detailed teacher's guide that describes classroom procedure, lists lab materials (mostly inexpensive household items), and includes reproducible forms for student use; and a handbook for students that guides project activities. The best way to teach the process of science is to have your students participate in the process.

Not all programs definitively answer questions presented. Those that do, however, need to present valid conclusions. For example, the video *Lion, King of the Beasts?* poses this question: Is the lion really "king of the beasts"? The program explores the balance between predator and prey and examines techniques that prey animals use to keep from being eaten. In the end, the program concludes that "no animal is king"; instead, all are part of a complete tapestry of life.

*SB&F* reviewers are also asked to comment on the following:

1. Scope (completeness)
2. Quality of videography, graphics, color, and animation
3. Optical techniques used
4. Value when compared with similar titles
5. Value of supplements, if any

The beauty of video and software is that students can see images and hear sounds of things and places too remote, hidden, or dangerous to experience otherwise. For example, in *Continents Adrift,* viewers can see the effects of plate tectonics by viewing the program's dramatic footage of the 1989 Loma Prieta earthquake and the spectacular 1980 eruption of Mount St. Helens. And, although few students will ever peer firsthand into the human body, with the *My Amazing Human Body* software program, they can "dissect" the human body, put parts of the body together from front and back views, and examine sections of the body by measuring, poking, spinning, and x-raying them.

The choice of optical techniques and how well they are used may also be determining factors in your purchasing decision. *Nagasaki Journal,* for example, combines recently discovered footage shot by Marines during their occupation of Nagasaki with photos taken the day after the atomic blast by Japanese Army photographer Yosuki Yamahata. Also, some techniques that work well for adults may not work well for young children, and vice versa. For example, *The Chariot Races: A Journey from Disabled to Enabled,* because of its artistic yet sometimes choppy switches from color to black and white (in flashbacks and so on), would be appropriate for middle school and older students, but not so well suited for younger students.

Supplements to a video or software program can add to or detract from the value of a purchase. *Continents Adrift: An Introduction to Continental Drift and Plate Tectonics* comes with a useful teacher's guide that includes learning objectives, a summary of the program, review and discussion questions, suggested activities, a glossary, a bibliography, and a script. *A Journey through the Cell* provides activity sheets that include pre-viewing questions, post-viewing questions, extended learning questions, hidden-word puzzles, diagrams of animal and plant cells, and an extensive glossary. Some supplementary materials, however, are not useful at all, adding nothing or even detracting from the program. For this reason, *SB&F* reviewers are asked to carefully evaluate all materials that accompany a program.

It is also important to determine if the material is equally appropriate for boys and girls, for various ethnic groups, and for people of various abilities. *SB&F* reviewers address the following equity considerations in their reviews:

Where appropriate, does the material have equal male/female representation? Is the text gender neutral? In hands-on science books, are girls as well as boys pictured performing the experiments?

Are females and minority groups portrayed in a nonstereotypical fashion?

Do illustrations, photographs, or film footage include examples of minority group members, senior citizens, or people with disabilities?

Would the material be relevant in a wide number of settings (that is, urban, suburban, and rural) and to a wide spectrum of students (that is, those who are economically disadvantaged as well as those who are wealthy)?

In hands-on resources, do the suggested experiments use materials that are accessible to economically disadvantaged students?

Would the material be especially useful in a multicultural curriculum?

Are the materials free of religious bias?

Do the materials provide a balanced presentation of controversial (for example, animal rights) or sensitive (for example, AIDS) issues? Are dissenting opinions presented fairly?

Although most materials these days use gender-neutral language, it is still critical to look at how males and females are portrayed in media. Often, a video will include equal numbers of boys and girls, but show the boys in more active roles, and the girls watching passively on the sidelines. The programs in the bibliography, however, have been evaluated using the *SB&F* equity criteria, and many have received high marks for their positive portrayals of females, minority group members, and people with disabilities. *The Crime Laboratory,* for example, shows grade-school boys and girls of various races solving the crimes. In the Bill Nye programs, children, adults, and scientists are representative of various ethnic groups, and gender equity is evident in all the presentations.

Although most of the programs listed in the bibliography include females and minorities, some specifically focus on them. The series *Breakthrough: The Changing Face of Science in America,* for example, profiles contemporary African American, Latino, and Native American scientists and engineers. Although these scientists had backgrounds that were diverse and, on occasion, difficult, all were able to set priorities, take risks, make a difference, and—not least—enjoy their work. Three of the videos listed—*Breathing Lessons: The Life and Work of Mark O'Brien, The Chariot Races: A Journey from Disabled to Enabled,* and *I'm Not Disabled*—tell the stories of people with disabilities.

Many producers take advantage of the unique quality of video to evoke emotional responses. Therefore, it is important to make sure that controversial or sensitive issues are balanced and that dissenting opinions are presented fairly. *AIDS and Your World,* for example, empowers young viewers to decide for themselves where to "draw the line" and avoid behaviors that may place them at risk of acquiring the infection. *Banking Our Genes* covers the application of DNA research to help identify violent criminals as well as the remains of servicemen and servicewomen. The video also discusses the vast array of tests that are being developed for diseases when the patient is presymptomatic. The potential advantages of these applications and tests are discussed, along with the potential dangers of allowing genetic information to be used wrongly. The series *Integrity in Scientific Research,* developed to address ethical questions that might face researchers, presents scenarios in shades of gray rather than in black and white, leaving the viewer with an open-ended, provocative situation that invites discussion.

## References

American Association for the Advancement of Science. *Resources for Science Literacy: Professional Development.* CD-ROM. New York: Oxford University Pr., 1997.

National Science Resources Center. *Resources for Teaching Elementary School Science.* Illustrated. Washington, D.C.: National Academy Pr., 1996. xv+288pp. Index.

———. *Resources for Teaching Middle School Science.* Illustrated. Washington, D.C.: National Academy Pr., 1998. xv+479pp. Index.

The *National Science Education Standards* includes a rating system for books, videos, software, and other reference materials. The codes are based on the level of difficulty, and will assist teacher and librarians in determining the intended audience for the reference materials. The following bibliography includes these codes:

K Preschool or kindergarten
EP Elementary, grades 1 & 2
EI Elementary, grades 3 & 4
EA Elementary, grades 5 & 6
JH Junior high, grades 7 & 8
YA Young adult, grades 9–12
C College
T Teaching professional
GA General audience

## Bibliography

### Science as Inquiry

#### K–4

*Learning Math and Science Together.* Hampshire College, School of Natural Science, Amherst, MA 01002. 1995. Color. 20 min. Video: $50. *SB&F* 32, no. 3, p. 88; T.

The video follows a group of teachers through the woods on a four-week summer field exercise that exposes them to new perspectives on learning and hands-on interactions with science.

*Science, Discovery, and Laughter: A Fun and Practical Guide for Doing Science with Children.* The Science Club, 55 First Place NW, Suite 4, Issaquah, WA 98027. 1994. Producer: Williams Communications. Color. 58 min. Activity book. Video: $19.95. *SB&F* 32, no. 8, p. 245; K–JH, T, GA.

This video and guidebook present hands-on science activities, exploring seven different science topics with at least two experiments each that can be performed both at school and at home.

**5–8**

*Science Scenarios*. Series. Meridian Learning Solutions, 236 East Front Street, Bloomington, IL 61701. 1997. Color. Teacher's guide and student research handbook. Video: $149.95 (each). EI, EA.

*The Consumer Research Center.* 12 min. *SB&F* 33, no. 7, p. 214.

This is a real-life learning kit targeted at grades 4–6. Each of the three modules is a self-contained unit developed as an extended activity lasting from eight to ten days. The kit consists of a research handbook for each student, a teacher's guide, and a videotape on the nature of the scientific work the students are to conduct. The intent is to have the students get directly involved in assessing products through a hands-on activity that follows the tenets of the scientific method.

*The Crime Laboratory.* 15 min. *SB&F* 33, no. 7, p. 214.

Three crimes involving theft or some other "soft" transgression are investigated. The teacher's guide includes additional assignments: three more thefts and a car accident. Interdependence among both individuals and teams is achieved by giving only one set of materials to each team, by having different teams do different tests on the evidence presented, and by requiring all the teams' results to solve the crime.

*The Geology Station.* 14:30 min. *SB&F* 33, no. 5, p. 148.

Working in teams, students gather, analyze, and interpret geological data on an undeveloped parcel of land. Nine one-hour class sessions direct general training, project orientation, and the formation of student teams; lab testing; the presentation of lab results; the construction of geological maps; the preparation of descriptive brochures; and the presentation of scientific results aimed at appropriate land development.

*The Health Organization.* 12:30 min. *SB&F* 33, no. 4, p. 118.

Of the nine hour-long sessions provided, two are devoted to orientation and basic training; three to collecting, compiling, and graphing the results of physical measurements; one to making scientific presentations; and the final three to planning, doing, and presenting the results of simple laboratory food testing.

*Spinning Things* and *Earthquakes*. Disney Educational Productions, 501 S. Cherry, Suite 350, Denver, CO 80246. 1996. From the *Bill Nye Sampler II Curriculum Set*. Color. 52 min. Video: $199 (series, 5 programs); $49.95 (each). *SB&F* 33, no. 3, p. 87; EA–YA.

*Spinning Things* explores the basic physics of rotating bodies, using examples from a perfect football pass to the rotation of the earth. Simple experiments anyone can do and humorous demonstrations draw attention to the basic science. *Earthquakes* begins with a simulated temblor and all the excitement and destruction that accompany it.

### 9–12

*Science Fair Projects: The Ultimate Guide*. Cambridge Educational, P.O. Box 2153, Charleston, WV 25328-2153. 1995. Color. 30 min. Video: $89. *SB&F* 32, no. 7, p. 214; YA, T, GA.

This video uses the comments of students, teachers, and judges to provide an overview of the science fair concept as well as a rationale for student participation.

## Physical Science

### 5–8

*The Convection of Heat*. Films for the Humanities and Sciences, Inc., Box 2053, 743 Alexander Road, Princeton, NJ 08540. 1996. From the *Eureka!* series. Color. 30 min. Video: $129; rental: $75. *SB&F* 33, no. 7, p. 215; JH.

This video moves smoothly through six major segments on volume and density, buoyancy, convection, heat as energy, radiation waves, and the radiation spectrum. The animation is humorous and colorful, and the content of the video is accurate.

*Physics: Why Bother?* PBS Video, 1320 Braddock Place, Alexandria, VA 22314-1698. 1995. Producer: IMG. Color. 60 min. $39.95. *SB&F* 32, no. 2, p. 55; JH–C, GA.

This video, featuring physicist Leon Lederman, successfully addresses several important concepts of atomic and nuclear physics in terms of easily comprehensible human functions and natural scenes.

*Solids, Liquids and Gases.* Rainbow Educational Media, Inc., 4540 Preslyn Drive, Raleigh, NC 27616. 1994. Producer: Cochran Communications. Color. 22 min. Teacher's guide. Video: $89. *SB&F* 32, no. 6, p. 185; EA, JH.

*Solids, Liquids, and Gases* is designed to give middle school students a basic understanding of the differences and similarities among these three phases of matter.

*What Is Chemistry?* Clearvue-EAV, 6465 N. Avondale Avenue, Chicago, IL 60631-1996. 1997. From the *What Is Science?* series. Color. 24:55 min. Teacher's guide. Video: $430 (series, 5 programs); $90 (each). Close-captioned. *SB&F* 33, no. 8, p. 247; JH, YA.

This short video includes most fields of chemistry and can lead to further discussion and exposition of each field. The teacher's guide is well written and helpful.

**9–12**

*The Chemistry Set.* Facts on File, Inc., 11 Penn Plaza, New York, NY 10001-2006. 1997. $299.95. System requirements: 486-16; Windows 3.1; 8 MB RAM; 2 MB free hard-disk space; SVGA 65K (16-bit true color) monitor; printer; mouse; double-speed CD-ROM drive; and sound card. *SB&F* 34, no. 2, p. 56; YA, GA.

*The Chemistry Set* is an interactive, enhanced version of the periodic table.

*Halliday CD-Physics 2.0.* John Wiley and Sons, 605 Third Avenue, New York, NY 10158. 1997. System requirements: PC: Windows 3.1 or later; 486; 33 MHz; 15 MB hard-disk space; 8 MB RAM; 640 x 480 pixels; 256 colors; MPC CD-ROM drive; sound card; and speakers. Macintosh: System 7.0; Series II or better; 15 MB hard-disk space; 8 MB RAM; 640 x 480 pixels; 256 colors; and double-speed CD-ROM drive. *SB&F* 34, no. 1, p. 25; YA, C.

This CD-ROM is based on the popular textbook by David Halliday, Robert Resnick, and Jearl Walker titled *Fundamentals of Physics,* fifth edition, published by Wiley. The program contains all the text of the printed book, but includes additional invaluable information in a "Student's Solutions Manual and Study Guide," which is a tutorial and review of the subject matter, along with examples and explanations not included in the print version of the text.

*The Power of Sound.* BBC Worldwide Americas, Inc., 747 Third Avenue, New York, NY 10017. 1997. Color. 30 min. Teacher's guide. Video: $59.95. *SB&F* 33, no. 9, p. 279; YA, C.

The strength of this film is its focus on the applications of ultrasound. After a brief introduction to sound, a wide variety of uses of ultrasonic sound are illustrated.

## Life Science

### K–4

*Amazing Animals.* Dorling Kindersley Multimedia, 95 Madison Avenue, 10th Floor, New York, NY 10016-7801. 1997. $29.95. System requirements: PC: 486DX; 33 MHz or higher processor; Windows 3.1x, 95, or later; 8 MB RAM (12 MB recommended for Windows 95); double-speed CD-ROM drive; sound card; mouse; SVGA 256-color display; and loudspeaker or headphones. Macintosh: System 7.0+; 68LC040; 25 MHz; 8 MB RAM (12 MB required for System 7.5+ and for PowerPC); 256 colors; double-speed CD-ROM drive; mouse; and loudspeaker or headphones. *SB&F* 33, no. 7, p. 217; K–EI.

With this software, students can learn about a variety of animals (armored and dangerous), animal babies, and animals that can disguise themselves.

*The Amazing Coral Reef* and *Coral Reefs: Rainforests of the Sea.*The Video Project, 200 Estates Drive, Ben Lomond, CA 95005. 1996. Color. 20 min. Teacher's guide. Video: $59.95; rental: $35. *SB&F* 33, no. 4, p. 116; EP–EA.

These two videos describe coral reefs and the dangerous forces that are marshaled against their survival worldwide. A considerable amount of time is given to the abundant life forms that coral reefs support—the reason coral reefs are sometimes referred to as the "rain forests of the sea."

*Armored Animals.* Dorling Kindersley Multimedia, 95 Madison Avenue, 10th Floor, New York, NY 10016-7801. 1996. From the *Amazing Animals* series. Color. 30 min. Video: $12.95. Close-captioned. *SB&F* 33, no. 5, p. 149; EP.

In this film, animals with armor, such as spines, thick skin, and shells, are shown in excellent sequences and explained by a narrator and a cartoon lizard named Henry. The definition of "armor" is broad—it includes elephant seal blubber and snake skin—but the examples are good.

*Come Walk with Me.* The Video Project, 5332 College Avenue, Suite 101, Oakland, CA 94618. 1994. Producer: Earthtalk Studios. Color. 28 min. Video: $59.95; rental: $30. *SB&F* 32, no. 5, p. 148; K–EI.

In this film, "Walkin" Jim Stolz, wilderness songster and long-distance hiker, takes a group of ten children on a day hike in the mountains of Montana to show them wilderness plants and animals in their natural habitats.

*Creatures of the Blue* and *Creatures of the Wild.* IVN Communications, Inc., 2246 Camino Ramon, San Ramon, CA 94583. 1995. From the *Sierra Club Kids* series. 29 min. Video: $14.95. *SB&F* 32, no. 9, p. 279; K–EA.

Set to pleasing music, these videos provide children from preschool to grade 6 an entertaining glimpse of a variety of animals.

*Kratts' Creatures.* Series. PBS Video, 1320 Braddock Place, Alexandria, VA 22314-1698. 1996. Color. 60 min. Video: $39.95 (each):

*In Search of the Tasmanian Tiger* and *Lion, King of the Beasts? SB&F* 33, no. 3, p. 87; EP–EA, GA.

In Search of the Tasmanian Tiger has the Kratt brothers retracing a route taken through Tasmania early in the century. Along the way, they encounter various examples of Tasmanian wildlife: snakes, dingoes, ring-tailed opossum, unusual birds, spiders, and the Tasmanian devil. Lion, King of the Beasts? has the Kratt brothers somewhere in the African savanna to answer the question: Is the lion really "king of the beasts?" As with other programs in this excellent wildlife series, the learning takes places through the use of awesome documentary footage, humor, and creative animation.

*Pan Troglodytes: An In-Depth Analysis* and *The Great Canadians. SB&F* 33, no. 3, p. 88; EP–EA, GA.

In Pan Troglodytes: An In-Depth Analysis, the Kratt brothers are teaching a trio of young orphan chimpanzees that are being prepared to return to the wild in Africa. First exploring the ancestral time line for primates, the program then briefly shows us different members of the ape family. In The Great Canadians, the Kratt brothers are roaming the Great Canadian North country as early settlers did—by canoe. The program introduces us to an abundance of diverse wildlife, including beavers, moose, caribou, bears, otters, loons, and elk.

*My Amazing Human Body.* Dorling Kindersley Multimedia, 95 Madison Avenue, 10th Floor, New York, NY 10016-7801. 1997. $29.95. System requirements: PC: 486DX2; 66 MHz; Windows 95; 26 MB hard-disk space; 12 MB RAM; double-speed CD-ROM; 8-bit sound card; mouse; 640 x 480 pixels; 256 colors; loudspeakers or headphones; and, to access online features, Internet access. Macintosh: System 7.1.2+; 68LC040; 33 MHz; 8 MB RAM (12 MB for System 7.5.3+ and PowerPC); 6 MB hard space; and all the rest as for PC. *SB&F* 34, no. 1, p. 25; EP, EI.

The four sections of this program—"Take Me Apart," "What Am I Made Of," "Build Me a Body," and "Me and My Day"—enable children to "dissect" the human body and put parts of the body together from front and back views. Users can examine sections of the body by measuring, poking, spinning, and x-raying them, and can take care of a body as it lives through a typical day.

**5–8**

*Alaska: Grizzly Country.* Chip Taylor Communications, 15 Spollett Drive, Derry, NH 03038-5728. 1996. From the *Environmental Studies* series. Color. 30 min. Video: $99.99. *SB&F* 33, no. 4, p. 116; EA–YA, GA.

*Alaska, Grizzly Country* provides the viewer the opportunity to see spectacular scenes of Alaska's vast wilderness. The film should be an especially valuable resource for middle and secondary school teachers whose objective is to teach students how to enjoy, preserve, and protect the environment.

*Audubon Society's Butterflies for Beginners.* MasterVision, 969 Park Avenue, New York, NY 10028. 1996. Color. 60 min. Video: $19.95. *SB&F* 32, no. 4, p. 118; JH–T, GA.

Focusing on thirty-two common butterfly species in North America, this video follows butterflies through their life cycle and gives instructions on how to raise a collection of butterflies.

*Birds.* National Geographic Society, Educational Services, 17th & M Streets, NW, Washington, DC 20036. 1997. From the *Animal Classes* series. Color. 22 min. Video: $99. Close-captioned. *SB&F* 33, no. 9, p. 279; EA.

In this outstanding video, a student tells his teacher and his class what a bird is and how birds differ from other animals. The role of each type of feather is described in the video. The program also covers features of the skeleton, lungs and air sacs, beaks, feet, and eyes, as well as nest building, feeding behavior, flying, and migration.

*Cell Biology: Membranes, Cell Motility.* Clearvue-EAV, 6465 N. Avondale Avenue, Chicago, IL 60631-1996. 1995. $225 (lab pack of 5 CD-ROMs); $75 (each). Set of 2: $420 (lab pack); $140 (each). System requirements: Macintosh: 68030 or better; Color QuickDraw (256 colors recommended); System 7 or later; and 6-MB RAM. Windows: 386/20-MHz processor; Windows 3.1 or later; 4 MB RAM; and VGA 640 x 480 with 16 colors. MS-DOS: 386/20 MHz processor; DOS 5.0 or later; 4 MB RAM; and VGA 640 x 480 with 16 colors. (MS-DOS version does not support sound play.) *SB&F* 32, no. 5, p. 152; JH.

This multimedia educational presentation explores cell membranes and motility.

*Cells and Genes.* The Mona Group, 23 Plumtree Road, P.O. Box 407, Sunderland, MA 01375-0407. 1997. $21.95. System requirements: PC: 486 Pentium processor; Windows 3.1 or 95; 16 MB; 4X or better CD-ROM drive; super VGA monitor; and sound card recommended. Macintosh: 68040 processor; 7.0 or better; 16 MB; 4X or better CD-ROM drive; color monitor; 8-bit at 640 x 480 pixels; and 16- or 24-bit recommended. *SB&F* 33, no. 9, p. 281; EA, JH.

*Cells and Genes* is a cornucopia of interesting, informative, and accurate presentations, emphasizing genetics and cellular biology and associated areas.

*The Crabs, the Birds, the Bay.* Bullfrog Films, P.O. Box 149, Oley, PA 19547. 1997. Producer: Natural Arts Films. Color. 19 min. Study guide. Video: $175; rental: $45. *SB&F* 34, no. 3, p. 85; EA–T, GA.

This film is a fascinating treatment of the relationship that exists on Delaware Bay between horseshoe crabs and migrating shorebirds early each spring.

*Dolphins.* National Film Board of Canada, 350 Fifth Avenue, Suite 4820, New York, NY 10118. 1998. From the *Champions of the Wild* series. Color. 25 min. Discussion guide. Video: $999 (series, 13 programs); $499 (any 6 programs); $129.95 (each). *SB&F* 34, no. 9, p. 275; EI–YA, GA.

This video describes the dolphin research of Diane Claridge and Ken Balcomb in the Bahama Islands. The researchers examine what happens to these animals when they are released into the wild after they have been confined and hand-fed for long periods.

*Eyewitness Encyclopedia of Nature 2.0.* Dorling Kindersley Multimedia, 95 Madison Avenue, 10th Floor, New York, NY 10016-7801. 1997. $39.95. System requirements: PC: Windows 3.1x/95; 486DX/33 MHz; 8 MB RAM; 22 MB available; 640 x 480 pixels; 256 colors (16-bit colors preferred); double-speed CD-ROM; 8-bit sound card; loudspeakers or headphones; mouse; and Internet access. Macintosh: System 7.0+; 68LCo40 25 MHz; 12 MB; 640 x 480 pixels; 256 colors (thousands of colors preferred); double-speed CD-ROM; 25 MB recommended available; 8-bit sound card; loudspeakers or headphones; mouse; and Internet access. *SB&F* 33, no. 9, p. 281; EI–T, GA.

*Eyewitness Encyclopedia of Nature 2.0* encompasses a wide variety of topics. In addition to extensive and accurate information on various types of plants and animals, which may be accessed by habitat or classification, the CD has sections on climate, natural history, and microscopic organisms.

*Fish* and *Marine Mammals, Mammals and Birds,* and *Wetlands and Rivers and Streams.* Disney Educational Productions, 501 S. Cherry, Suite 350, Denver, CO 80246. 1996. From the *Bill Nye Sampler II Curriculum Set.* Color. 52 min. Video: $199 (series, 5 programs); $49.95 (each). *SB&F* 33, no. 4, p. 117; EA, JH.

These excellent videos emphasize major concepts and processes of science in an entertaining, humorous, and often unique manner.

They are also consistent with national standards and goals in science education.

*Insects*. National Geographic Society, Educational Services, 17th & M Streets, NW, Washington, DC 20036. 1997. From the *Animal Classes* series. Color. 22 min. Teacher's guide. Video: $99. Close-captioned. *SB&F* 33, no. 9, p. 279; EA.

This video is a one-man show on the position of insects in the living world: A sixth-grade student narrates the excellent insect photography as a class project. Remarks by other students in the class are interspersed.

*In the Wild*. Series. PBS Video, 1320 Braddock Place, Alexandria, VA 22314-1698. 1995. Producer: Tigress Productions. Color. 60 min. Video: $100 (series); $39.95 (each).

*Dolphins with Robin Williams* and *Gray Whales with Christopher Reeve*. *SB&F* 32, no. 4, p. 119; EI–T.

These two videos take viewers on an adventure with Williams and Reeve as they watch the featured animals in their natural habitats and learn about their struggle for survival.

*Pandas with Debra Winger*. *SB&F* 32, no. 5, p. 149; JH, YA, GA.

Actress Debra Winger and her eight-year-old son Noah travel through China to visit the giant panda and discuss its chances of survival.

*A Journey through the Cell, Part One. Cells: An Introduction*. Cambridge Educational, P.O. Box 2153, Charleston, WV 25328-2153. 1996. Color. 25 min. Teacher's guide. Video: $69.95. *SB&F* 33, no. 4, p. 118; JH, YA.

This video introduces students to cells and their properties. Each cell organelle is shown, its function described, and its relationship to other parts of the cell delineated.

*A Journey through the Cell, Part Two. Cell Functions: A Closer Look*. Cambridge Educational, P.O. Box 2153, Charleston, WV 25328-2153. 1996. Color. 25 min. Teacher's guide. Video: $69.95. *SB&F* 33, no. 4, p. 118; JH, YA.

This video, which introduces students to energy storage and release, protein synthesis, and cell reproduction, is an excellent introduction to concepts that often are difficult for the beginning biology student.

*Legends of the Blue.* IVN Communications, Inc., 2246 Camino Ramon, San Ramon, CA 94583. 1997. From the *Wild Animals* series. Color. 60 min. Video: $24.99. *SB&F* 33, no. 7, p. 216; JH–T, GA.

This fascinating video presents legends of wild animals that live in lakes, rivers, and the oceans. The legends range from scary to gentle, and many of the animals shown are unfamiliar to our Western culture.

*Mountain in the Sea and Filming Secrets.* PBS Video, 1320 Braddock Place, Alexandria, VA 22314-1698. 1997. Producer: Howard Hall Productions. From the *Secrets of the Ocean Realm* series. Color. 60 min. Video: $150 (series, 5 programs); $39.95 (each). Close-captioned. *SB&F* 34, no. 6, p. 181; JH–C, GA.

This is a beautifully produced journey to the aquatic world of Cocos Island, 300 miles off the west coast of Costa Rica. *Mountain in the Sea and Filming Secrets* captures marine life beautifully on film, showing dramatic and rare footage of hammerhead sharks being cleaned by barberfish, stingrays moving en masse, and frogfish in a most unusual courting ritual. The video also provides insight into the filmmaking process—the tedious preparation, massive cameras, specialized diving gear, and cumbersome movie lights—that makes the images possible.

*Mystery of the Senses.* Series. PBS Video, 1320 Braddock Place, Alexandria, VA 22314-1698. 1995. Producer: Green Umbrella Ltd. for WETA. Color. 60 min. Video: $79.95. *SB&F* 32, no. 4, p. 119; JH, YA, GA.

*Hearing*

Hosted by Diane Ackerman, *Hearing* is a series of vignettes depicting how the auditory sense works.

*Smell*

Focuses on those aspects of chemical communication picked up by the olfactory system and then interpreted by the brain.

*National Audubon Society Interactive CD-ROM Guide to North American Birds.* Alfred A. Knopf, 201 E. 50th St., New York, NY 10022; 800-793-2665. 1996. $56.95. System requirements: PC: 25-MHz 386 dx or faster; Windows 3.1 or 95; DOS 5.0; 8 MB RAM; VGA/SVGA monitor for 640 x 480 resolution; 256-color monitor; mouse; 12 MB memory; MPC-compatible CD-ROM drive with at least 150 KB transfer rate; Sound Blaster Pro or compatible. Macintosh: Powerbook 180c or higher; Macintosh II, Quadra, or Performa series; Motorola 68030 processor; 640 x 480 resolution; 256-color monitor; System 7.0 or higher; 5 MB RAM; QuickTime 2.0 or higher (provided); and CD-ROM drive (double-speed recommended). *SB&F* 32, no. 6, p. 186; EI–T, GA.

This CD-ROM field guide includes: (1) an ecology section with landscape photography and "soundscape" events that awaken pictures with environmental life; (2) a section on 723 bird species, with over 2,000 color photographs and training in identifying birds that involves size, shape, color, sound, location, and life zones with maps and five video essays; and (3) a notebook section that is structured for the creation of lists for bird watching, authoritative information on species, personal tests for developing identification skills, and proposed planning trips to twenty "hot spots" for seeing select species.

*Nocturnal Animals: The Night Shift.* National Geographic Society, Educational Services, 17th & M Streets, NW, Washington, DC 20036. 1997. Color. 22 min. Teacher's guide. Video: $99. Close-captioned. *SB&F* 33, no. 8, p. 249; JH, YA, GA.

*Nocturnal Animals* is an amusing, interesting film that mixes comical black-and-white film clips with fabulous videos of animals in action. The accompanying teacher's guide is concise and adequate, and includes objectives, vocabulary words, and activities.

*Oceans: Earth's Last Frontier.* Rainbow Educational Media, Inc., 4540 Preslyn Drive, Raleigh, NC 27616. 1995. Producer: Cochran Communications. Color. 24 min. Teacher's guide. Video: $89. *SB&F* 32, no. 7, p. 215; EI–T, GA.

This beautifully designed and informative video shows a student class aboard a ship that is part of the Coastal Ecology Learning Program on Long Island Sound. The video gives general information concerning the causes of ocean waves and currents, the formation of the sea floor, and marine plant and animal life and their interaction.

*Rediscovering the Amazon.* Lucerne Media, 37 Ground Pine Road, Morris Plains, NJ 07950. 1996. Color. 23:57 min. Video: $195; rental: $60. *SB&F* 33, no. 8, p. 248; EI–YA.

*Rediscovering the Amazon* is a well-done film with a nice emphasis on people, cultures, and the role of the Amazon basin. The film raises many questions about conflicts between cultures, the maintenance of tribal systems and beliefs, and the true value of the area.

*Science in Your Own Backyard.* Series. Lucerne Media, 37 Ground Pine Road, Morris Plains, NJ 07950. 1995. Producer: Digital Nutshell, Inc. Color. Video: $395 (series, 3 programs); $145 (each).

*Praying Mantis.* 10 min. *SB&F* 32, no. 7, p. 216; EA, JH.

Using ear-catching music, this video features the life cycle of the praying mantis, showing the natural progression from egg to nymph to adult.

*The Private World of Jean Henri Fabré.* 10 min. *SB&F* 32, no. 7, p. 216; EA, JH, GA.

Through the explorations of French entomologist Fabré, viewers experience the human side of science, as Fabré's love for observing a variety of insects is matched by his ability to describe them in ways that make language soar.

*Spiders.* 22 min. *SB&F* 32, no. 7, p. 216; EI–YA.

Beginning with the myths of ancient Greece, this film leads viewers through the wide-ranging world of spiders.

*Tallgrass Prairie: An American Story.* National Geographic Society, Educational Services, 17th & M Streets, NW, Washington, DC 20036. 1997. Color. 28 min. Video: $99. Close-captioned. *SB&F* 33, no. 7, p. 215; EI–YA.

*Tallgrass Prairie* depicts the dynamic forces in an ecosystem and how these forces interact. The impact of bison, climate, wildfire, and people on this unusual ecosystem demonstrates how species can either survive through cooperation and sharing or be destroyed through thoughtless competition.

*Terrarium.* BFA Educational Media, 2349 Chaffee Drive, St. Louis, MO 63146. 1997. Color. 12 min. Discussion guide. Video: $240. *SB&F* 33, no. 7, p. 215; EI–JH.

This video demonstrates how to make a terrarium and shows many different types of terraria. Despite the film's brevity, there are references to the water cycle, recycling, and the humane treatment of animals, each of which could be useful in stimulating discussion.

*The Ultimate 3D Skeleton.* Dorling Kindersley Multimedia, 95 Madison Avenue, 10th Floor, New York, NY 10016-7801. 1996. $29.95. System requirements: Multimedia PC 486DX; 33MHz or higher processor; 4 MB RAM; MPC-compatible CD-ROM drive; sound card; mouse; SVGA 640 x 480 256-color display; and loudspeakers or headphones and Microsoft Windows 3.1x or Windows 95. *SB&F* 32, no. 9, p. 281; JH–T, GA.

This interactive CD-ROM is well suited for audiences of all ages who are interested in learning the structure of the human skeleton.

*The Wonderful World of the Butterfly.* Lucerne Media, 37 Ground Pine Road, Morris Plains, NJ 07950. 1994. Color. 30 min. Video: $195; rental: $60. *SB&F* 32, no. 1, p. 22; EI–YA, GA.

This video provides an overview of the biology of butterflies; describes their life cycle, evolution, and so on; and discusses the need to preserve butterfly species.

## 9–12

*Bird* and *Cat.* Dorling Kindersley Multimedia, 95 Madison Avenue, 10th Floor, New York, NY 10016-7801. 1996. From the *Eyewitness Virtual Reality* Series. $49.95. System requirements: PC 486SX/25-MHz or higher processor; 4 MB RAM; MPC-compatible CD-ROM drive; sound card; mouse; SVGA 256-color display; and loudspeakers or headphones. *SB&F* 33, no. 2, p. 56; YA, GA.

*Bird* and *Cat* provide detailed information, suitable for general interest and courses in life science, biology, and zoology, and related areas.

*Genetics CD-ROM.* Clearvue-EAV, 6465 N. Avondale Avenue, Chicago, IL 60631-1996. 1995. $225 (lab pack of 5 CD-ROMs); $75 (each). System requirements: Macintosh: 68030 or better; Color QuickDraw (256 colors recommended); System 7 or later; and 6 MB RAM. Windows: 386/20-MHz processor; Windows 3.1 or later; 4 MB RAM; and VGA 640 x 480 with 16 colors. MS-DOS:

386/20-MHz processor; DOS 5.0 or later; 4 MB RAM; and VGA 640 x 480 with 16 colors. (MS-DOS version does not support sound play.) *SB&F* 32, no. 5, p. 152; YA–T, GA.

Parts I and II of *Genetics* discuss simple Mendelian genetics, including segregation and independent assortment, from a "historical" perspective, using the work of Mendel as a paradigm. Part III uses the work of Griffith (1928), Avery, MacLeod, and McCarty (1944); and Hershey and Chase (1952); and work on the tobacco mosaic virus to convince us that genes are made of nucleic acid.

*Introduction to Cells*. Human Relations Media, 175 Tompkins Avenue, Pleasantville, NY 10570. 1997. Producer: Cochran Communications. Color. 22:16 min. Teacher's guide. Video: $189. *SB&F* 34, no. 4, p. 118; YA.

This video is an extremely well done presentation of cell organization from the structure-function aspect. The narrator uses Prospect Park, in Brooklyn, New York, as a focal point to emphasize the ubiquitousness of cells in biology.

*Life Is Impossible*. BBC Worldwide Americas, Inc., 747 Third Avenue, New York, NY 10017. 1997. Color. 50 min. Teacher's guide. Video: $34.95. *SB&F* 33, no. 7, p. 215; YA, C.

*Life Is Impossible* begins with the contention that life is not scientifically possible and proceeds to explore many of the current theories of the origin of life on earth. The presentations should provide the viewer with sufficient information to allow for an evaluation of the various competing hypotheses for the origin of life.

*Peterson Multimedia Guides: North American Birds*. Houghton Mifflin Interactive, 120 Beacon Street, Somerville, MA 02143. 1996. $69.95. System requirements: 486 DX; 8 MB RAM; Windows 3.1 or higher; SVGA display; double-speed CD-ROM drive; MPC-compatible sound card (22 KHz, 16 bit); 20 MB hard-drive space; and mouse. For Internet access option: 14.4 KBPS modem and 6 MB hard-disk space. *SB&F* 32, no. 9, p. 280; GA.

This multimedia CD-ROM contains the text, artwork, and range maps of the current *Peterson Field Guide: Eastern Birds and Western Birds,* plus an array of additional information for studying all bird species encountered in North America.

*The Smallest Organisms.* Films for the Humanities and Sciences, Inc., Box 2053, 743 Alexander Road, Princeton, NJ 08540. 1997. From the *World of Living Organisms* series. Color. 15 min. Video: $59.95. *SB&F* 33, no. 7, p. 215; YA, GA.

*The Smallest Organisms* presents a very general overview of bacteria and viruses. There is a brief discussion of the properties of these microbes that separate them from animals and plants. The video also includes brief discussions of how the immune responses of the body function against disease agents and of selected aspects of genetic engineering and gene therapy.

*Working with Whales.* Lucerne Media, 37 Ground Pine Road, Morris Plains, NJ 07950. 1997. Color. 18 min. Video: $195; rental: $60. *SB&F* 34, no. 4, p. 118; YA, C.

Captive marine mammals are the focus of commercial marine parks and public aquaria. This video contrasts whales that are held in captivity for purely entertainment purposes with those that are maintained for educational and scientific reasons.

## Earth and Space Science

### K–4

*Weather Station.* BFA Educational Media, 2349 Chaffee Drive, St. Louis, MO 63146. 1997. From the *Backyard Science* series. Color. 12 min. Discussion guide. Video: $240. *SB&F* 33, no. 8, p. 248; EP–EA.

Using simple and easily understood terms, this video describes the many instruments found in a weather station and explains the function of each instrument well. The first part of the video carefully explains how weather instruments are used to gather data and the second half shows students how to create a weather station.

### 5–8

*Continents Adrift: An Introduction to Continental Drift and Plate Tectonics.* Rainbow Educational Media, Inc., 4540 Preslyn Drive, Raleigh, NC 27616. 1995. Producer: Peter Matulavich Productions. Color. 27 min. Teacher's guide. Video: $99. *SB&F* 32, no. 8, p. 245; EA, JH.

This video introduces middle school students to plate tectonics through outstanding photography, instructive animation, simple yet effective classroom demonstrations, and easy-to-follow narration.

*Earthquakes: Our Restless Planet.* Rainbow Educational Media, Inc., 4540 Preslyn Drive, Raleigh, NC 27616. 1995. Producer: Peter Matulavich Productions. Color. 22 min. Teacher's guide. Video: $99. *SB&F* 32, no. 8, p. 246; EA, JH.

*Earthquakes* introduces students to the relationship between plate tectonics and seismic activity.

*Earthquakes: Quake, Rattle and Roll; Plate Tectonics: Solving the Puzzle;* and *Volcanoes: Fire from Within.* Lucerne Media, 37 Ground Pine Road, Morris Plains, NJ 07950. 1995. Producer: Double Diamond Corp. Color. 20 min. Video: $145; rental: $60. *SB&F* 32, no. 7, p. 214; EA–YA.

These three videos, with clear and accurate commentary and illustrated with attractive visuals, present three major topics of interest to geologists and the public today.

*The Earth Science CD-ROM.* Clearvue-EAV, 6465 N. Avondale Avenue, Chicago, IL 60631-1996. 1995. $180 (lab pack of 5 CD-ROMs); $60 (each). System requirements: Macintosh: 68030 or better; Color QuickDraw (256 colors recommended); System 7 or later; and 6 MB RAM. Windows: 386/20-MHz processor; Windows 3.1 or later; 4 MB RAM; and VGA 640 x 480 with 16 colors. MS-DOS: 386/20-MHz processor; DOS 5.0 or later; 4 MB RAM; and VGA 640 x 480 with 16 colors. (MS-DOS version does not support sound play.) *SB&F* 32, no. 7, p. 218; EA, JH.

The program is divided into six sections—on rocks and minerals, oceans, mountains, deserts, air and the atmosphere, and rain and clouds—that are presented as mysteries being solved by a Sherlock Holmes type of detective.

*Earth's Natural Resources CD-ROM.* Society for Visual Education, Inc., 6677 North Northwest Highway, Chicago, IL 60631. 1996. $225 (lab pack of 5); $75 (each). System requirements: Macintosh: 68030 processor or better; 8 MB RAM; System 7.0 or later; and 256-color display or better. Windows: 386/33-MHz processor or better; 8 MB RAM; Windows 3.1 or later; and 16-color VGA monitor with 640 x 480 resolution (SVGA with 256 colors recommended). *SB&F* 33, no. 5, p. 151; EA, JH, GA.

*Earth's Natural Resources* consists of modules titled "Air, Water, and Soil Resources," "Wildlife Resources," "Energy Resources," and "Future Resources."

*The Grand Tour of Voyager 2: Jupiter, Saturn, Uranus, Neptune.* BFA Educational Media, 2349 Chaffee Drive, St. Louis, MO 63146. 1996. From the *Science Spectrum* series. Color. 26 min. Discussion guide. Video: $2,495 (series, 13 programs); $225 (each). *SB&F* 33, no. 6, p. 181; JH, YA.

Using computer animation and photographs collected by the spacecraft itself, this video depicts *Voyager 2*'s encounters with the planets Jupiter, Saturn, Uranus, and Neptune, and their associated moons.

*What Is Earth Science?* Clearvue-EAV, 6465 N. Avondale Avenue, Chicago, IL 60631-1996. 1997. From the *What Is Science?* series. Color. 28:39 min. Teacher's guide. Video: $430 (series, 5 programs); $90 (each). Close-captioned. *SB&F* 34, no. 5, p. 149; JH–T, GA.

This film describes the main branches of earth science, including meteorology, geology, and oceanography. Excellent live-action photography of the natural processes that occur in each field is included.

*Why Are the Mountains So High? An Introduction to Mountains and Mountain Building.* Rainbow Educational Media, Inc., 4540 Preslyn Drive, Raleigh, NC 27616; 1995. Producer: Peter Matulavich Productions. Color. 26 min. Teacher's guide. Video: $99. *SB&F* 32, no. 8, p. 246; EA, JH.

This video introduces students to mountain building and its relationship to plate tectonics, focusing on three basic types of mountains: fold, volcanic, and fault block.

**9–12**

*The Earth CD-ROM.* Clearvue-EAV, 6465 N. Avondale Avenue, Chicago, IL 60631-1996. 1995. $180 (lab pack of 5 CD-ROMs); $60 (each). System requirements: Macintosh: 68030 or better; Color QuickDraw (256 colors recommended); System 7 or later; and 6 MB RAM. Windows: 386/20-MHz processor; Windows 3.1 or later; 4 MB RAM; and VGA 640 x 480 with 16 colors. MS-DOS:

386/20-MHz processor; DOS 5.0 or later; 4 MB RAM; and VGA 640 x 480 with 16 colors. (MS-DOS version does not support sound play.) *SB&F* 32, no. 7, p. 218; YA.

This multimedia CD-ROM presentation is divided into four sections: an introduction and segments on the composition of the earth, geologic processes, and the history of the earth.

*The Search for Alien Worlds.* Warner Home Video, 4000 Warner Boulevard, Burbank, CA 91522. 1997. Producer: Engel Brothers Media, Inc., and Thomas Lucas Productions, in association with Devillier Donegan Enterprises and PBS. From the *Mysteries of Deep Space* series. Color. 60 min. Video: $49.95 (series, 3 programs). *SB&F* 34, no. 1, p. 24; GA.

This video explores humanity's continuing interest in the possibility of extraterrestrial life.

*To the Edge of the Universe.* Warner Home Video, 4000 Warner Boulevard, Burbank, CA 91522. 1997. From the *Mysteries of Deep Space* series. Color. 60 min. Video: $49.98 (series, 3 programs). *SB&F* 33, no. 9, p. 279; GA.

From amateur astronomers attending a "star party" in Texas to professionals using the Hubble Space Telescope to explore the limits of the universe, the video shows the lure of astronomy from many viewpoints. The graphics and animation are of high quality and very colorful.

*VolcanoScapes V: Hawaii Volcanoes National Park, an Historical Perspective.* Tropical Visions Video, Inc., 62 Halaulani Place, Hilo, HI 96720. 1996. Color. 60 min. Video: $29.95. *SB&F* 33, no. 1, p. 22; GA.

*VolcanoScapes V* is a skillful blending of spectacular footage depicting historic eruptions at Hawaii Volcanoes National Park within both cultural and scientific contexts.

## Science and Technology

### K–4

*Bumptz Science Carnival.* Theatrix Interactive, 1250 45th Street, Suite 150, Emeryville, CA 94608-2924; 510-658-2800. 1995. ISBN 1-887661-01-8. System requirements: Windows: PC 486/25-MHz or better; 8 MB RAM; 256-color VGA; double-speed CD-ROM drive; MS-DOS 5.0; Windows 3.1 or Windows 95; 16-bit sound card; mouse; and speakers. Macintosh: Macintosh LCIII or better; 8 MB RAM; 256-color monitor; double-speed CD-ROM drive; and System 7.0. *SB&F* 32, no. 3, p. 89; K–EI.

This program combines a carnival-like atmosphere; interesting graphics that will undoubtedly grab a child's attention; a number of games; solutions to the games, which require critical thinking; educational cartoons, which cursorily explain force, magnetism, light, density, viscosity, and a number of other subjects; and experiments to be done away from the computer.

*Peanut Butter: How It's Made.* Lerner Media, 241 First Avenue North, Minneapolis, MN 55401. 1996. System requirements: Macintosh: 13-inch monitor or larger; 256 colors; double-speed CD-ROM drive; Macintosh LCIII or better; 8 MB RAM (4 MB minimum); and System 7.0 or better with multimedia support. *SB&F* 33, no. 6, p. 185; EP–EA.

Fasten your seat belts and take an educational ride along the various roads surrounding "Lernerville" to discover how peanuts are raised, harvested, marketed, and made into America's favorite food: peanut butter.

**5–8**

*The Digital Environment.* The Video Project, 5332 College Avenue, Suite 101, Oakland, CA 94618. 1995. Produced in association with the California Academy of Sciences. From the *Kids from CAOS* series. Color. 29 min. Video: $199 (series, 4 programs); $59.95 (each). *SB&F* 32, no. 7, p. 213; EA–YA.

This video exposes students of all ages to the use of computers in science.

*Energy Choices.* MediCinema Ltd., 131 Albany Avenue, Toronto, ON M5R 3C5. 1995. From the *Energy, the Pulse of Life* series. Color. 43 min. Teacher's guide. Video: $350. *SB&F* 32, no. 6, p. 184; EA–YA.

Part 2 of the award-winning series *Energy, the Pulse of Life, Energy Choices* contains both Canadian and U.S. examples of energy use and development.

*Garbage into Gold.* The Video Project, 5332 College Avenue, Suite 101, Oakland, CA 94618. 1995. Color. 25 min. Video: $39.95; rental: $25. *SB&F* 32, no. 5, p. 147; EA–T, GA.

This video confirms the distance we have come in nationwide recycling efforts and provides proof of the significant rewards that effort has produced.

*Good Garbage.* Cintia Cabib, 4242 East-West Highway, #712, Chevy Chase, MD 20815. 1995. Color. 19 min. Discussion guide. Video: $55. *SB&F* 32, no. 3, p. 88; EI–C, GA.

*Good Garbage* follows recyclable garbage items from the time they are discarded to their deposition in landfills and incinerators and as they are recycled.

*Inventions/Computers.* Disney Educational Productions, 501 S. Cherry, Suite 350, Denver, CO 80246. 1997. From the *Bill Nye the Science Guy* series. Color. 46:18 min. Teacher's guide. Video: $49.95. Close-captioned. *SB&F* 34, no. 5, p. 150; EP–YA.

The first segment of the tape deals with commonplace inventions, ranging from the chocolate chip cookie to the light bulb. The second segment of the tape is a high-quality explanation of the hows and whys of computers.

*Microchips with Everything.* Lucerne Media, 37 Ground Pine Road, Morris Plains, NJ 07950. 1995. Producer: Scottish Television for ITV. From the *Science and Society* series. Color. 21 min. Student activities guide. Video: $195; rental: $60. *SB&F* 32, no. 7, p. 213; EA, JH.

From Glasgow traffic control to virtual reality to a "smart card," this snappy introduction to the ubiquitous computer features computer science professors who give cogent views of the significance and pitfalls of computer usage.

*New York Underground.* PBS Video, 1320 Braddock Place, Alexandria, VA 22314-1698. 1997. From the *American Experience* series. Color. 60 min. Video: $59.95; home video version: $19.98. Close-captioned. *SB&F* 33, no. 7, p. 216; JH–T, GA.

*New York Underground* is an excellent portrayal of one of the world's largest projects, undertaken at a time when new technologies were changing the world.

*The Peace of Paper: An Introduction to Origami.* World Information Videos, Inc., 310 Riverside Drive, #200B, New York, NY 10025. 1997. Color. 53 min. Video: $79.95. *SB&F* 33, no. 9, p. 280; EA–YA, T, GA.

*The Peace of Paper* is essentially an instructional videotape that introduces the art and craft of paper folding known as origami. The viewer feels encouraged to work along with the artists demonstrating their techniques on film. The video includes origami demonstrations by adults and a child who represent culturally diverse backgrounds.

*Simple Things You Can Do to Save Energy in School.* The Video Project, 5332 College Avenue, Suite 101, Oakland, CA 94618. 1995. Producer: The Noodlehead Network, in association with Burlington Electric. Color. 15 min. Video: $49.95; rental: $25. *SB&F* 32, no. 5, p. 147; EA–YA.

This video presents a "how-to" guide for students to help conserve energy in their schools.

*The Wright Stuff.* PBS Video, 1320 Braddock Place, Alexandria, VA 22314-1698. 1996. From the *American Experience* series. Color. 60 min. Video: $69.95. *SB&F* 32, no. 8, p. 246; GA.

Garrison Keillor narrates the story of the Wright brothers—their setbacks and successes, their personalities and passions, and their fame and subsequent battles against those who sought to steal the accolades due the inventors of the airplane.

**9–12**

*Nanotopia.* BBC Worldwide Americas, Inc., 747 Third Avenue, New York, NY 10017. 1997. Color. 48 min. Teacher's guide. Video: $54.95. *SB&F* 33, no. 8, p. 249; YA.

*Nanotopia* describes the progress scientists and engineers have made toward building machines atom by atom and explores the awesome potential of a world in which these machines are a reality. The video comes with an excellent teacher's supplement, including an exercise with microscopes that will give students experience with the tiny size of the machines described and pictured.

*No Time to Waste: Resource Conservation for Hospitals.* Fanlight Productions, 47 Halifax Street, Boston, MA 02130. 1995. Producer: Ann Carol Grossman and Ben Achtenberg. Color. 30 min. Study guide and leader's manual. Video: $250; rental: $50. *SB&F* 32, no. 6, p. 184; YA, C, GA.

This video is a detailed account of how hospitals reduce environmentally harmful wastes and deal with the problem of waste in hospital management.

## Science in Personal and Social Perspectives

### K–4

*Groark Learns about Bullying; Groark Learns about Prejudice; Groark Learns to Control Anger; Groark Learns to Listen;* and *Groark Learns to Work Out Conflicts.* Live Wire Media, 3450 Sacramento Street, San Francisco, CA 94118. 1995. From the *Prevent Violence with Groark* series. Color. 28 min. Discussion guide. Video: $299.50 (series, 5 programs); $69.95 (each). *SB&F* 32, no. 6, p. 183; EP, EI.

In these videos, children interact with the puppet Groark as he helps them learn how to get along with others.

### 5–8

*The Chariot Races: A Journey from Disabled to Enabled.* Live Wire Media, 3450 Sacramento Street, San Francisco, CA 94118. 1997. Color. 30 min. Video: $69.95. Close-captioned. *SB&F* 33, no. 7, p. 213; EA–T, GA.

This excellently photographed film features a highly athletic young man who becomes a paraplegic. He describes his feelings and his return to high-level competitive downhill sports using a unique, high-tech, off-road wheelchair and, subsequently, a "mono ski."

*Food for Thought.* Lucerne Media, 37 Ground Pine Road, Morris Plains, NJ 07950. 1995. Producer: Scottish Television for ITV. From the *Science and Society* series. Color. 21 min. Student activities guide. Video: $195; rental: $60. *SB&F* 32, no. 4, p. 120; JH, YA, GA.

This educational video focuses on many controversial food safety and food production issues and delivers a balanced presentation on the dangers and benefits of bacteria, fungi, molds, and yeasts.

*Home Place*. Series. Bullfrog Films, P.O. Box 149, Oley, PA 19547. 1998. Producer: Waterhen Film Productions. Color. Video: $495 (series, 4 programs); $195 (each). Rental: $125 (series); $45 (each). *SB&F* 34, no. 8, p. 246; EA–T, GA.

*Going Home*

Explores the conflicts humans have with the natural ecosystems that sustain life.

*Inside-Outside*

Explores the problems of seeing things from one's own perspective, as opposed to viewing them from a completely different perspective that is neither self-centered nor brief.

*Life Cycles*

Presents the calcium cycle to show viewers that all life on earth is linked by a delicate ecosystem that has eco-organisms among its many parts.

*Partnership*

Posits the thesis that the main reason species become extinct on our planet is the disappearance of natural habitat. The video gives an example of the attempt to save the Swiss fox in the newly established Grasslands National Park in Saskatchewan, Canada.

*Making a Difference*. The Video Project, 5332 College Avenue, Suite 101, Oakland, CA 94618. 1995. Produced in association with the California Academy of Sciences. From the *Kids from CAOS* series. Color. 29 min. Video: $199 (series, 4 programs); $59.95 (each). *SB&F* 32, no. 7, p. 213; EA–YA.

In this video, students are shown how they and others can help to alleviate many of the threats to individual species and the environment.

*Nagasaki Journal*. The Video Project, 5332 College Avenue, Suite 101, Oakland, CA 94618. 1995. Producer: Independent Documentary Group, in association with BBC2. Color. 28 min. Video: $79; rental: $45. *SB&F* 32, no. 4, p. 120; JH–T, GA.

This film by Emmy Award–winning filmmakers shows how the atomic bomb shaped the lives and thoughts of two Japanese survivors and a U.S. Marine, one of the first Americans to arrive in the bombed city.

*Planet Neighborhood.* Series. Bullfrog Films, P.O. Box 149, Oley, PA 19547. 1997. Producer: WETA. Color. 56 min. Teacher's guide or viewer's guide. Video: $395 (series, 3 programs); $195 (each). Rental: $150 (series, 3 programs); $75 (each). Close-captioned. *SB&F* 34, no. 4, p. 116; JH–T, GA.

*Community*

Covers (1) community action; (2) alternative energy; (3) water purity and the challenge to achieve it; (4) new biotechnology in Burlington, Vermont's "Living Machine," which uses natural means to purify water; and (5) Chattanooga's efforts to create a cleaner and more sustainable city.

*Home*

Addresses (1) the retrofitting of an old home to make it more energy efficient and more comfortable; (2) the Super Efficient Refrigerator Program (SERP), which designs refrigerators so that they lower ozone depletion and use less energy; (3) the horizontal-axis washing machine, which uses a third of the water of conventional vertical-axis machines; (4) windows that are relatively leak proof and heat proof; (5) the construction of homes with less wood and with alternative materials, such as steel, fly ash, and scrap metal, making the homes less costly to construct; and (6) community composting and smart shopping.

*Work*

Examines (1) indoor air quality, as found in a building with a serious mold problem, and how it affects the health of those working in the building; (2) the Durst Building in New York City (the first skyscraper to be green, or environmentally sensitive); (3) industrial ecology, as exemplified by Union Carbide in Sea Drift, Texas, a company acting to purify its emissions and reduce waste; and (4) recycling parts of cars and alternative models of car engines.

*Taking Charge of Me: Emotional I.Q.* Sunburst Communications, Inc., 101 Castleton St., Pleasantville, NY 10570-0100. 1998. Color. 26 min. Teacher's guide. Video: $99.95. *SB&F* 34, no. 5, p. 148; JH, YA.

This video presents teens in conflict and suggests methods for dealing with their problems.

*Talking about Sex: A Guide for Families.* Planned Parenthood Federation of America, Inc., 810 Seventh Avenue, New York, NY 10019. 1996. Producer: Buzzco Associates. Color. 30 min. Resource guide for parents and activity book. Video: $29.95. *SB&F* 32, no. 7, p. 214; EA–YA, T, GA.

*Talking about Sex* presents a wonderfully visual, musical, and very sensible introduction to the topic of body changes that accompany growing up.

*Tina's Journal.* The Video Project, 5332 College Avenue, Suite 101, Oakland, CA 94618. 1995. Producer: The San Francisco Recycling Program. Color. 17 min. Video: $65; rental: $35. *SB&F* 32, no. 5, p. 148; EA–YA.

In this video, viewers follow Tina in her exploration of the sources and destinations of household waste.

**9–12**

*AIDS and Your World.* Fanlight Productions, 47 Halifax Street, Boston, MA 02130. 1996. Color. 25 min. Video: $195; rental: $50. *SB&F* 33, no. 2, p. 56; YA, C.

This video brings the universal message of HIV prevention to teens and college-age students through interviews with individuals infected with HIV or diagnosed with AIDS, breaking the stereotypical view of these people.

*The Amazon, Part I,* and *The Amazon, Part II.* Benchmark Media, 569 North State Road, Briarcliff Manor, NY 10510. 1997. Producer: Christian Science Monitor. From the *Environmental Studies* series. Color. 27 min. Teacher's guide. Video: $395; rental: $60. *SB&F* 34, no. 3, p. 84; YA–T.

These educational videos provide a dramatic story of the opposing pressures of the Brazilian economic crisis and its impact on the Amazon river and rain forest.

*Amazon Journal.* Filmakers Library, 124 East 40th Street, New York, NY 10016. 1995. Producer: Realis Pictures, Inc. Color. 58 min. Video: $395; rental: $75. *SB&F* 32, no. 6, p. 183; YA–T, GA.

This video focuses on the native peoples of the Amazon and their struggles to retain their way of life in the face of gold miners, loggers, and the local Brazilian government.

*Banking Our Genes.* Fanlight Productions, 47 Halifax Street, Boston, MA 02130. 1996. Color. 33 min. Video: $145; rental: $50. *SB&F* 32, no. 6, p. 186; YA–T.

This video gives an excellent overview of the excitement and issues facing us as we move into the new genetic era.

*Breathing Lessons: The Life and Work of Mark O'Brien.* Fanlight Productions, 47 Halifax Street, Boston, MA 02130. 1996. Producer: Inscrutable Films. Color. 35 min. Video: $195; rental: $50. *SB&F* 32, no. 9, p. 279; C, GA.

Poet-journalist Mark O'Brien, who was severely disabled with polio as a young child and now spends most of his life in an iron lung, narrates his own story in this award-winning video.

*BSE for Teens.* American Journal of Nursing Company, 555 West 57th Street, New York, NY 10019-2961. 1996. Color. 7 min. Video: $200; rental: $95. *SB&F* 32, no. 7, p. 216; YA–T, GA.

This short but instructive video on breast self-examination (BSE) is introduced by Jennie Garth of "Beverly Hills 90210," who urges teenage girls to begin BSE as a lifelong healthy habit.

*Cry of the Forgotten Land.* The Video Project, 5332 College Avenue, Suite 101, Oakland, CA 94618. 1995. Producer: The Endangered Peoples Project and Gryphon Productions. Color. 26 min. Video: $89; rental: $45. *SB&F* 32, no. 9, p. 278; YA–T, GA.

*Cry of the Forgotten Land* is directed at increasing public awareness of the plight of the native Moi people of Indonesia, who have occupied the rain forests of the western tip of New Guinea for thousands of years.

*The Fourth "R": Responsibility.* Sunburst Communications, Inc., 101 Castleton Street, Pleasantville, NY 10570-0100. 1998. Color. 24 min. Teacher's guide. Video: $99.95. *SB&F* 34, no. 5, p. 148; YA.

This tape, a teacher's edition, presents four situations in which students have to deal with decisions requiring the acceptance of responsibility for their own actions.

*Global Warming*. Hawkhill Associates, Inc., 125 E. Gilman Street, Madison, WI 53703. 1998. Color. 31 min. Teacher's guide. Video: $79. *SB&F* 34, no. 7, p. 215; YA–C, GA.

This collection of the views of several luminaries from well-known institutions gives an excellent perspective on the complexity of the issues and the range of opinions concerning global climate.

*I'm Not Disabled*. Landmark Media, Inc., 3450 Slade Run Drive, Falls Church, VA 22042. 1995. Color. 24 min. Video: $195. *SB&F* 32, no. 1, p. 23; YA–T, GA.

This German-made film is about disability from the perspective of four people who use sports and physical activity for rehabilitation as well as increased interaction and control of their environment.

*Nukes in Space: The Nuclearization and Weaponization of the Heavens, Parts 1 and 2*. Chip Taylor Communications, 15 Spollett Drive, Derry, NH 03038. 1995. Producer: EnviroVideo. From the *Environmental Studies* series. Color. 52 min. Video: $100. *SB&F* 32, no. 5, p. 149; YA–T, GA.

In this exposé of a potentially significant environmental threat, viewers learn how the U.S. government already has employed, or plans to employ, plutonium-powered nuclear technology in space.

*Patient Education Video*. Series. Milner-Fenwick, Inc., 2125 Greenspring Drive, Timonium, MD 21093. 1996. Color. 10 min. Video: $1,149 (series, 14 programs); $89 (each). Available in Spanish. Close-captioned:

*Healthy Food Choices: Daily Decision Making*. *SB&F* 33, no. 6, p. 183; YA–T, GA.

Faced with a potentially life-threatening illness, such as adult-onset diabetes, anyone who has the ability to make choices that affect the progression of the disease would begin with the rudiments of daily decision making: If food is (in part) my medicine, how do I make intelligent selections?

*Healthy Food Choices: Developing a Plan*. *SB&F* 33, no. 6, p. 184; YA–T, GA.

Adult-onset diabetes is one of many disorders that can be strongly influenced by dietary management. From this superbly produced teaching video comes a summary of meal-planning basics for everyone, even though the thrust of meal planning in the video is for those whose blood sugar needs stabilizing.

*Monitoring Your Blood Sugar: Key Concepts*. *SB&F* 33, no. 6, p. 184; YA–T, GA.

This video covers a key concept in controlling diabetes: how to monitor and maintain one's level of blood sugar within a "target range," which varies for each person, depending on daily variables, such as level of exercise, amount and quality of food eaten, stress factors, and medications taken.

*Promise and Perils of Biotechnology: Genetic Testing*. Cold Spring Harbor Laboratory Press, 10 Skyline Drive, Plainview, NY 11803-2500. 1996. Color. 25 min. Teacher's guide. Video: $70. *SB&F* 33, no. 3, p. 88; YA, C.

This video dramatically portrays some implications of genetic testing using the true-life stories of people faced with the prospects of Huntington's disease or familial hypercholesterolemia.

*Risky Business: Biotechnology and Agriculture*. Bullfrog Films, P.O. Box 149, Oley, PA 19547. 1996. Color. 25 min. Video: $195; rental: $45. *SB&F* 33, no. 5, p. 150; YA–T, GA.

This impressive production deals with agricultural biotechnology and the controversial issue of genetic testing. In a sentence, genetic engineering involves transferring the DNA of one animal or plant species into another to achieve certain desirable properties, such as resistance to disease. This can be a very difficult subject to understand, but the presentation in *Risky Business* is quite straightforward, avoiding much of the biochemical jargon without sacrificing scientific accuracy, and focusing on the effects and consequences of such practices.

*Secrets of the Choco*. Bullfrog Films, P.O. Box 149, Oley, PA 19547. 1997. Color. 52 min. Video: $250; rental: $75. *SB&F* 33, no. 7, p. 213; C, T, GA.

This award-winning film documents a tragedy beginning to happen. The Choco is the narrow belt of tropical rain forest between Colombia's Pacific coast and its Andes Mountains. When the filmmakers visited the area in 1995, it was relatively pristine; the viewer is allowed to sample the remarkable diversity of flora and fauna. But the sad fate of all such places seems inevitable.

*Sex and Other Matters of Life and Death*. The Cinema Guild, 1697 Broadway, Suite 506, New York, NY 10019. 1995. Color. 60 min. Video: $295; rental: $95. *SB&F* 32, no. 9, p. 280; YA, T, GA.

This video presents a powerful, dramatically stated plea for frank and full presentation of the public health consequences of unsafe sex.

*Sex, Teens, and Public Schools*. Filmakers Library, 124 East 40th Street, New York, NY 10016. 1995. Producer: Public Policy Productions. Color. 55 min. Video: $395; rental: $75. *SB&F* 32, no. 4, p. 117; JH, YA, T, GA.

Set in the real world as a documentary, this video offers various, expectedly controversial approaches to dealing with teen sexual activity.

## History and Nature of Science

### K–4

*Back to School: How Scientists and Engineers Can Help Schools*. Research Triangle Park Chapter of Sigma Xi, P.O. Box 13068, Research Triangle Park, NC 27709. 1995. Color. 120 min. For availability, contact Sigma Xi. *SB&F* 32, no. 9, p. 279; T, GA.

This documentation of a public forum provides a wealth of ideas, suggestions, and concrete examples of how K–8 classrooms can benefit from visiting scientists and engineers.

*Muttaburrasaurus: A Dinosaur of Gondwanda*. BFA Educational Media, 2349 Chaffee Drive, St. Louis, MO 63146. 1997. Color. 26 min. Discussion guide. Video: $225. *SB&F* 33, no. 8, p. 248; EP–JH.

This delightful video combines animation with real-life scientists at work, giving students an opportunity to see the world of this newly discovered dinosaur through sight and sound.

## 5–8

*Astronomy and Art.* Landmark Media, Inc., 3450 Slade Run Drive, Falls Church, VA 22042. 1994. Color. 28 min. Video: $225. *SB&F* 32, no. 5, p. 148; JH, YA, T.

From this video, viewers will gain important insights into the function of artwork in communicating the latest scientific theories and discoveries.

*The Boyhood of John Muir.* Bullfrog Films, P.O. Box 149, Oley, PA 19547. 1997. Producer: Florentine Films. Color. 78 min. Video: $275; rental: $85. *SB&F* 34, no. 5, p. 151; EI–C, T, GA.

This dramatization accurately portrays Muir's years on the Wisconsin frontier, beginning with the family's emigration from Scotland in 1849 and concluding a decade later with the terrible, nearly blinding accident that so profoundly altered the course of American environmental history.

*Charles Drew: Determined to Succeed.* Churchill Media, 6901 Woodley Avenue, Van Nuys, CA 91406-4844. 1995. From the *African American Achievers* series. Color. 30 min. Discussion guide. Video: $149.95. *SB&F* 32, no. 1, p. 23; JH–T, GA.

This biography highlights an outstanding African American scientist whose contributions included the development of blood banks and the use of blood plasma during World War II.

*David Brower: A Conversation with Scott Simon.* Bullfrog Films, P.O. Box 149, Oley, PA 19547. 1997. Producer: John DeGraaf with KCTS TV/Seattle. Color. 56 min. Video: $195; rental: $75. *SB&F* 34, no. 2, p. 53; JH–T, GA.

This video is an introduction to the environmental movement and to David Brower, longtime Sierra Club director, founder of Friends of the Earth, and a key leader of the environmental movement.

*Keeping the Earth: Religious and Scientific Perspectives on the Environment.* Union of Concerned Scientists, Two Brattle Square, Cambridge, MA 02238. 1996. Color. 27 min. Discussion guide. Video: $14.95. *SB&F* 33, no. 2, p. 54; EP–T, GA.

*Keeping the Earth* is a religious and scientific exchange about the human impact on the environment and on all created life that will be meaningful for all ages. The discussion guide is very well done and is essential for making good use of the video.

*Making It Happen: Masters of Invention.* Churchill Media, 6901 Woodley Avenue, Van Nuys, CA 91406-4844. 1995. Color. 22 min. Discussion guide. Video: $149.95. *SB&F* 32, no. 5, p. 150; EA–C, GA.

Using period prints and archival film, this video features devices invented or enhanced by free and slave African Americans (e.g., Benjamin Banneker, Elijah McCoy, Patricia Bath) and used worldwide.

*A Naturalist in the Rainforest: A Portrait of Alexander Skutch.* Bullfrog Films, Oley, PA 19547. 1995. Producer: Paul Feyling. Color. 54 min. Video: $250; rental: $80. *SB&F* 32, no. 5, p. 151; JH–T, GA.

This film portrays the life and achievements of Skutch, who devoted more than fifty years of research to understanding avian life in tropical habitats.

*One Doctor: Daniel Hale Williams.* The Cinema Guild, 1697 Broadway, Suite 506, New York, NY 10019. 1997. Producer: Rex Barnett. Color. 43 min. Video: $99.95. *SB&F* 34, no. 5, p. 150; JH–C, GA.

*One Doctor: Daniel Hale Williams* introduces the viewer to "the father of black surgery," who earned that title by pioneering open-heart surgery and by establishing quality health care and nurse and physician training for African Americans.

**9–12**

*Breakthrough: The Changing Face of Science in America.* Series. PBS Video, 1320 Braddock Place, Alexandria, VA 22314-1698. 1996. Color. 60 min. Video: $250 (series, 6 programs); $49.95 (each):

*An Atmosphere of Change. SB&F* 33, no. 2, p. 54; YA, GA.

This video focuses on the backgrounds and careers of three environmental scientists—a marine ecologist whose research on toxic pollutants played a key role in protecting wildlife in San Francisco Bay and on the Aleutian Islands; a Nobel Prize–winning chemist who discovered the connection between spray can gases and ozone depletion; and an environmental toxicologist who first became concerned about damage to Native American communities by coal strip mining in Colorado.

*A Delicate Balance. SB&F* 33, no. 5, p. 147; JH–T, GA.

This video focuses on the importance of mentoring in shaping careers and on the difficulties and positive aspects of being an academic scientist. By portraying three talented and respected scientists, this video helps students explore options and understand problems academic scientists may face.

*Engineering from the Inside Out. SB&F* 33, no. 4, p. 118; JH–C.

This inspirational video features remarkable scientists and is a wonderful instrument for recruiting young, intelligent, underrepresented women and men into science. Especially good is the profile of Hector Medina, which presents a very practical application of the physics and architecture of the highly complex science of bridge building.

*The Path of Most Resistance. SB&F* 33, no. 2, p. 55; YA, GA.

This video describes the backgrounds and careers of four physicists. These scientists faced doubts, difficulties, and fears familiar to many young people, but their stories make clear that success in science depends much more on a love of adventure, an active imagination, and a willingness to take risks than on one's cultural, social, or economic background.

*Science and the American Dream. SB&F* 33, no. 2, p. 55; YA–T, GA.

This video profiles three scientist-engineers—their experiments and their struggles to incorporate the results of their work in the world of business. Included is a discussion of the social aspects of their research and some of the dilemmas faced by minority scientists.

*With Nerve and Muscle. SB&F* 33, no. 2, p. 55; EA–C.

Profiles three contemporary biologists who are making major contributions to our understanding of disease. The lives of these biologists (an African American studying nerve cells in squids, a Navajo Indian investigating how muscles form, and a woman from Puerto Rico studying dinoflagellate toxins) are presented from childhood to midlife as successful scientists.

*Edison's Miracle of Light.* PBS Video, 1320 Braddock Place, Alexandria, VA 22314-1698. 1995. From the *American Experience* series. Color. 60 min. Video: $69.95. *SB&F* 32, no. 6, p. 186; YA–T, GA.

This video presents an overview of Thomas Edison's background and some of the events that earned him the sobriquet "electrical wizard."

*Noah's Dilemma; Of Mice and Mendoza: Sharing in Science; Only a Bridge; Where Credit Is Due;* and *The Whole Truth.* American Association for the Advancement of Science, 1200 New York Avenue, NW, Washington, DC 20005. 1996. From the *Integrity in Scientific Research* series. Color. 9 min. Discussion and resource guide: $10.00. Video: $79.95. *SB&F* 32, no. 8, p. 244; YA–T, GA.

These videos address ethical questions that might face those involved in scientific research, such as: "Is it acceptable to use information gotten through 'back channels' to further one's own research?" and "To what extent should those who contribute to a scientific conclusion be given credit?"

*Talking with David Frost: Bill Gates.* PBS Video, 1320 Braddock Place, Alexandria, VA 22314-1698. 1995. Color. 60 min. Video: $69.95. *SB&F* 32, no. 4, p. 120; YA–T, GA.

In this hour-long conversation, viewers will not learn the secret to becoming rich, but will learn how the future looks to the world's most successful software entrepreneur.

# 5

# Using the Internet to Develop Science Literacy

MARIA SOSA

Educators and professional education organizations have developed or are developing national standards in mathematics, the arts, science, English, history, geography, foreign languages, civics, and physical education. Though these standards are voluntary, they establish high national benchmarks toward which all schools, teachers, and students can aspire. As these standards are being developed and refined, the Internet has simultaneously created an information explosion that can, and should, build on these articulated goals.

## Finding What Is Good

When the Internet first became popular, nearly all the information published on it came from academic sources and had clear scholarly value. As the technology of creating Web-based resources became more accessible, just about anyone could publish information on the Internet that *looked* professional and *sounded* credible. The truth is that anyone can say just about anything over the Internet. As more and more people gain the technological sophistication of Internet publishing, many more sites acquire the look and feel of authoritative content, so much so that they can fool all but the most astute users. These factors make it critical that websites be evaluated carefully for content and credibility.

There are a number of reliable sources you can use to find good websites. Scientific societies often create Internet resources for students. Some examples of these can be found in our list of resources in this chapter. These same organizations also provide links to other resources that are often, though not always, reviewed by knowledgeable staff. At the American Association for the Advancement of Science (AAAS), we have a Web resource called *Science NetLinks* (http:\\www. sciencenetlinks.org), which contains a wealth of resources divided broadly into two main sections. In "Science Websites," students can browse or search by keyword through our list of comprehensive "Super Science Sites" or our broader index of "Reviewed Science Sites." In our "Curriculum Connections" section, teachers will find an array of resources for bringing the science standards into the classroom. Also, *Science Books & Films* (SB&F) *Online* (http://www.sbfonline.com/) contains a section of annotated links called "Hot Links," which point users toward resources that contain good information on such topics as careers in science, the Human Genome Project, assisted technology, and more.

## Criteria for Evaluating Science Websites

AAAS is devising a procedure for reviewing World Wide Websites in the context of K–12 science, mathematics, and technology education. A two-part process is proposed for developing comprehensive reviews that can provide valuable information for a variety of users, including teachers, librarians, and students. The process involves gathering and verifying information from site submitters, then evaluating each site according to AAAS guidelines.

### Part One: Site Description

In the first part of the process, the following information would be requested from the site developer:

1. Target audience:
    - Teachers; students; public; science researchers

- Reading level: 4th grade; 6th grade; 8th grade; high school; college
- Level of science literacy: novice; college preparation; college; advanced

2. Goal of site creators:
   - To inform colleagues of work
   - To inform students, teachers or the public of work
   - To provide activities for students
   - To provide tools for students to connect science and policy information
   - To provide history of science connections
3. Target science topics:
   - Site submitter would select keywords for searching, probably from a list of topics that would correlate with 2061 benchmarks and science standards.
4. Description of additional documentation or materials available, if any (guidebooks, disks, and so on).
5. Suggestions for use of material for students, if any.
6. Anticipated context of use:
   - Science content for teachers
   - Science content for students doing projects
   - Hands-on labs for group or class
   - Scientific online discussions
   - Assessment of science understanding
   - Class science activity
7. Technical requirements: Multimedia, etc.
8. Human resources for FAQs (frequently asked questions) or on-line mentoring, if available, including evidence of expertise.
9. Date when site was first created, how often the site is revised, and the last revision date.
10. Reviews of the site, if known.
11. Information about links to related sites, including description of how these links are compiled and maintained.

## Part Two: Site Assessment

After the preceding information is compiled, reviewers would conduct independent evaluations of each site. Each reviewer would be asked to consider the following guidelines or criteria in the review:

1. Determine whether the site description as submitted by the developer is accurate. If not, the reviewer should complete or extend the site description, where appropriate.
2. Describe the breadth of science content and assess whether the science content is appropriate for: K–3; 3–6; 7–9; 10–12; adult; science teacher; specialist; etc.
3. Assess the credibility and relevance of science content for the target audience. For example,
   - Is the content credible and relevant for a novice (introduces understandable content connected to everyday experience)?
   - Is the content credible and relevant for students (complements textbook material)?
   - Is the content credible for researchers?
   - Is the content credible and relevant for the educated novice (enhances research in a different or related field)?
   - Is the content flawed or misleading for general public use?
4. Assess the pedagogical usefulness of the site, using the following criteria:
   - Does the site provide background for teachers?
   - Does the site augment textbook material for teachers or students?
   - Does the site provide evidence for scientific discussion, a project, or an activity?
   - Does the site stimulate a learner to think, investigate, or reflect by offering help, answers to questions, or probing questions?
   - Does the site provide tools for hands-on or simulated experimentation?
   - Does the site combine advertisements with information?
   - Does the site take advantage of technology for visualization, search, and so on?
5. Rate the clarity and organization of the site (excellent, good, poor).
6. Assess the amount of time required to review the site (too much; about right; very little).
7. Provide an overall rating (1 to 5 stars).

# Exploring Science Online

## Websites for Students

An interest in science is rooted in the natural world. Scientists formulate and test their explanations of nature using observations, experiments, and theoretical and mathematical models. It is important that students recognize that science is not just a collection of static facts and that scientific knowledge is subject to modification. Though it is best if children acquire this knowledge through direct experience and inquiry-based learning, the Internet can help students develop their natural curiosity and acquire the fundamental background needed for further learning.

The following websites range from the home pages of well-known institutions to those of lesser-known organizations, projects, or individuals. All of these sites, selected from *Science NetLinks* or *SB&F Online* recommendations, provide doorways to virtual places of science that students can explore.

### About Rainbows
**http://www.unidata.ucar.edu/staff/blynds/rnbw.html**

This site contains in-depth information on rainbows, including diagrams, quotes, text, labs, and directions for classroom demonstrations. It could be especially useful for middle-school science fair projects.

### Agropolis
**http://agcomwww.tamu.edu/agcom/agrotext/agcommap.html**

This Texas A&M University System Agriculture Program site contains beautiful photographs in the Digital Dragonfly section. Students can learn about agriculture in the Treasure Hunt game. Also fun is the interactive agriculture story, which weaves information provided by the user into a story.

### AllerDays
**http://www.allerdays.com/inkids.html**

Librarians, teachers, parents, and kids will find this site full of helpful information about allergies that the layperson can understand. It has good diagrams and clear explanations about what causes allergies and how the immune system responds.

**Amusement Park Physics**
http://www.learner.org/exhibits/parkphysics/

Part of the Annenberg/CPB Project Online Exhibits Collection, this site is designed to look at the "forces behind the fun."

**Arizona Science Center**
http://aztec.asu.edu/government/Tempe/asc/asc.html

A video tour of the museum as well as information about programs and activities at this brand new science center can be found at this site, which also provides links to current science hot topics.

**Asthma**
http://galen.med.virginia.edu/~smb4v/tutorials/asthma/asthma1.html

This site was developed especially for kids who have asthma and their families. The easy-to-read text, pictures, sounds, and movies help explain why asthma happens, what its symptoms are, how it is treated, and what happens to lungs during asthma attacks.

**Australian A–Z Animal Archive**
http://www.aaa.com.au/A_Z/

This alphabetical listing of most of Australia's native creatures describes each animal's range, dimensions, appearance, and eating and breeding habits, and indicates the family to which it belongs.

**The Bug Club**
http://www.ex.ac.uk/bugclub/

A club for children or the young at heart, the Bug Club is operated by students predominantly from the Biology Department at the University of Exeter in the United Kingdom. It includes an online newsletter and "care sheets" for taking care of pet cockroaches, stick insects, crickets, and tarantulas!

**The Carnegie Museum of Natural History's Discovery Room Online**
http://www.clpgh.org/cmnh/discovery/

Modeled after the Carnegie Museum's Discovery Room, this virtual-reality space allows visitors of all ages to touch and see things up close, explore our world, and learn something new.

**The Chicago Academy of Sciences**
http://www.chias.org/

This site contains a virtual tour of exhibits along with links to other regional resources and projects. Best of all, it is home to the Online Learning Center and Studio, a nifty educational cyber center for science learning.

**Chickadee Net**
http://www.owl.on.ca/chick/chick.html

Designed for children eight and under, this page is a spin-off from a children's nature magazine of the same title. It has a Tell Me Why section, along with puzzles and jokes.

**Children's Butterfly Site**
http://www.mesc.nbs.gov/butterfly.html

From the U.S. Geological Service Learning Web, this page contains answers to frequently asked questions, a picture gallery of websites, and much more.

**The Children's Museum of Indianapolis**
http://www.a1.com/children/home.html

Excellent online exhibits are the highlight of this high-quality website targeted to kids.

**Cool Hidden Stuff**
http://tommy.jsc.nasa.gov/~woodfill/SPACEED/SEHHTML/cool.html

This site contains answers to such questions as: How did the *Challenger* spacecraft explode? What famous comic strip characters traveled to outer space? Did you know that stars die?

**The Cub Den**
http://www.nature-net.com/bears/cubden.html

Even the youngest Web surfers will enjoy this site. Young children can hear a bear roar, read simple articles about bears, and look at pictures of a cub.

**Cyber Zoomobile**
http://www.primenet.com/~brendel/

Aimed largely at children, this database contains pictures of popular zoo animals divided into families with brief descriptions of each family.

**Dr. Bob's Interesting Science Stuff**
http://ny.frontiercomm.net/~bjenkin/science.htm

Interesting science tidbits for children age ten and up can be found here. Kids can even ask Dr. Bob questions or post their answers to questions in a section called The Fringe Forum.

**Dragonfly**
http://miavx1.muohio.edu/~Dragonfly/

The website for *Dragonfly,* the National Science Teachers Association's science journal for kids, contains an interactive feature article from the current issue and a selected article from previous issues. A great feature is the Dragonfly e-mail list, which connects children with real scientists and other young investigators around the world.

**Energy Quest**
http://www.energy.ca.gov/education/index2.html

This is a good site for grownups and kids. Kids can click on icons that lead them to pages on such topics as saving energy, super scientists, science projects, and alternative fuel vehicles. There is also a page of education resources for parents and teachers that includes descriptions of various energy education projects and links to science organizations and science sites.

**Entomology for Beginners**
http://www.bos.nl:80/homes/bijlmakers/ento/begin.html

This page gives some quick information about how bugs develop and reproduce. Users can click on different parts of a cute bug drawing to find out what the body part is called. The explanations are simple and effective for elementary school kids. The site also has links to more bug sites.

**Exploratorium**
http://www.exploratorium.edu

An online version of the famous science center, this site includes forums on hot topics, interactive activities, and a digital library. The Exploratorium also hosts the Institute for Inquiry and the Exploratorium Teacher Institute, both of which are featured on this site.

**Federation for Amateur Scientists**
http://web2.thesphere.com/SAS/

This site is the online home of the Amateur Scientists' Forum, an Internet networking service that links amateur and professional scientists. It includes a section that lists opportunities for amateur scientists and ideas for science fair projects. This site provides stimulating science information for motivated middle school and high school students.

**The Field Museum of Natural History**
http://rs6000.bvis.uic.edu:80/museum

At this site, kids can learn about what the world was like before the dinosaurs, and then learn about dinosaurs themselves. The site contains exhibits and information about the natural world and is updated frequently.

**Flight at the Children's Museum of Indianapolis**
http://www.a1.com/children/flyact.htm

Kids can learn about different types of flight and make their own flying devices, such as paper airplanes and origami birds. The site includes activities that can help users attract butterflies to the backyard and discover the wonder of flight.

**The Franklin Institute Science Museum**
http://sln.fi.edu/

Take a field trip without leaving home! Curious explorers can visit imaginary planets or go inside the human body to explore the heart. This site also contains links to lists of subtopics, science resources, and teacher resources available on the Internet, as well as links to other educational sites.

**Girl Tech**
http://www.girltech.com/HTMLworksheets/MS_mentors_home.html

This award-winning website has over 200 pages of fun educational content. It features games, an advice column, a weekly diary, women role models, science projects, inventions, and sports. It also has a boys' area designed to encourage and increase communication and understanding between girls and boys.

**Glacier**
http://www.glacier.rice.edu

For anyone who has ever wanted to learn more about Antarctica, this is a great place to begin. Both teachers and kids can use this site, which explores Antarctica and features live reports from the ice. The National Science Foundation and the Education Development Center in Boston developed the site.

**The Global Online Adventure Learning Site**
http://www.goals.com/

Explorers of all ages can participate in a growing number of exciting and educational adventures. The site includes virtual field trips, travel, adventure, science, technology, and nature.

**Gorillas**
http://www.selu.com/~bio/gorilla/

This site covers a wide variety of information about gorilla behavior, reproduction, and natural history. It has many photographs and a Natural History section that includes new information on the population and distribution of gorillas in the wild.

**Houston Museum of Natural Science**
http://www.hmns.mus.tx.us/

Take a look at stars in the planetarium, visit the Cockrell Butterfly Center, learn about insects and rocks, and lots more. There are about twenty different exhibits to explore.

**Imagination Station**
http://museums.mdmi.com/imaginationstation/

This is the website of a science center in Wilson, North Carolina, dedicated to the advancement of informal, inquiry-based science education.

**The Indianapolis Zoo**
http://www.indyzoo.com/

Along with information about museum programs, this website contains a virtual zoo and a kid's corner.

**The Information Dirt Road**
http://www.ics.uci.edu/~pazzani/4H/InfoDirtRoad.html

Brought to you by the 4-H Farm at the University of California, Irvine, this site allows users to download chicken sounds, read jokes about goat ears, and peruse articles by kids and their teachers on topics ranging from "Why raising a pig is fun" to rabbit labor and delivery.

**The Internet Webseum of Holography**
http://www.enter.net/~holostudio/

Here users will find hologames, an online exhibit gallery of holography, and even directions for making an inexpensive holography table for home and school from readily available materials. The site also has a holo-kids section, which contains a kid-friendly introduction to holography.

**Inventure Place**
http://www.invent.org/

This site contains an online book on the National Inventors Hall of Fame. Users can even nominate someone online for the Hall of Fame! The site also contains information about the B. F. Goodrich Collegiate Inventors Program, and a tour of the exhibits, some of which include audio clips.

**Izzy's Skylog**
http://darkstar.swsc.k12.ar.us/~kwhite/skylog.html

The basic purpose of this page is to promote and encourage interest in astronomy as a hobby. The site includes a Sky Calendar that shows prominent events for the month and a constellation of the month, and an Ask Izzy section in which user questions are answered.

**The Kansas City Museum**
http://www.kcmuseum.com/

Cute graphics and fun experiments are among the highlights of this science center site.

**Keith's Tomato Home Page**
http://www4.ncsu.edu/eos/users/k/kdmuelle/public/hp.html

Everything you ever wanted to know about tomatoes can be found on this page from Keith Mueller of North Carolina State University. Users can learn about starting seeds, cross-pollination, tomato varieties, diseases, and so on. This page is fun, friendly, and instructive.

**Kids Space**
http://www.kids-space.org

Kids under twelve can learn how to use the Internet and share their art, music, and stories with other kids. The site even goes to the trouble of explaining computer and Net terms like *browser, URL,* and *e-mail.* One particularly fun part is the site's StoryBook, a page where kids can read stories by other kids and even send in their own stories!

**KinderGarden**
http://aggie-horticulture.tamu.edu/kinder/index.html

This is a great site for teachers and kids. The Fun Page for Kids! has lots of activities kids can do to learn more about plants (for example, there is information on edible flowers, an ant farm, games, and project ideas).

**Lawrence Hall of Science**
http://www.lhs.berkeley.edu/

Kids who visit this site will find information about exhibits and programs, a comprehensive list of resources for teachers, a bat quiz, and much more.

**Liberty Science Center**
http://www.lsc.org/

Not only does this exciting website offer information about programs and exhibits, it is home to Cockroach World and The Yuckiest Site on the Web, where Wendell the Inquiring Worm will answer questions about things yucky or not.

**Liftoff Academy**
http://liftoff.msfc.nasa.gov/academy/academy.html

Shuttle missions are the focus of this site, which also contains information on topics ranging from the Russian space program to our universe. Take time to check out the Cool Facts. This page is appropriate for elementary students and older.

**Little Shop of Physics**
http://129.82.166.181/default.html

Looking for ideas for physics experiments? This site has a collection of hands-on science experiments designed for students at all grade levels. The experiments can be done at home and emphasize that kids can do interesting science with simple tools. The Amazing Physics of Everyday Objects page has experiments that can be done with stuff kids have around the house.

**The Mad Scientists Network**
http://medicine.wustl.edu/~ysp/MSN/MAD.SCI.html

The "network" provides a forum in which people can learn more about the world around them. The site has three primary divisions: (1) Ask-A-Scientist, which includes the online archive of questions and answers; (2) MAD Labs, which consists of hands-on science activities; and (3) MadSci Library, which includes links to other Ask-A-Scientist sites, and information about careers in science.

**Miami Museum of Science**
http://www.miamisci.org/

A highlight of this website is the hurricane storm center, an interactive exploration for kids. The site also features highly informative updates on the museum's exhibits and links to educational resources.

**Missouri Botanical Garden**
http://www.mobot.org/

There's so much information on this site that it's almost impossible to describe. It includes such treats as weekly updates of photos and information about plants that are in bloom. It includes a link to the Missouri Botanical Garden's MGnet for Kids and Teachers, a beautiful and comprehensive botany resource for kids.

**Monterey Bay Aquarium**
http://www.mbayaq.org

This site includes a gorgeous cyber tour of Monterey Bay; excellent educational resources, such as a shark word search; and information about research and conservation activities at the aquarium.

**The Muscle Page for Kids**
http://www.concentric.net/~Orthodoc/kidmuscle1.shtml

Lots of useful illustrations and easy-to-understand explanations make this a colorful and engaging site for kids to learn about the muscles in their bodies. The illustrations and photos show kids where muscles are located and explain how they work and what they do.

**The Museum of Science and Industry**
http://www.concentric.net/~Orthodoc/kidmuscle1.shtml

Anyone who has spent time in this Chicago museum will appreciate the comprehensive and educational virtual exhibits based on such popular attractions as the coal mine, the hatching chicks, and the U-505 submarine. Many of these online exhibits feature video and audio enhancements.

**The National Zoological Park**
http://www.si.edu/organiza/museums/zoo/nzphome.htm

This graphics-heavy site from the National Zoo features live webcams of zoo animals and pictures and information about zoo babies, as well as games and an enhanced photo library.

**Natural History Museum of Los Angeles**
http://www.lam.mus.ca.us/lacmnh

The website of this Los Angeles science center includes games for kids, research updates, and online exhibits.

**Neuroscience for Kids**
http://weber.u.washington.edu/~chudler/neurok.html

This site contains "brain basics" for kids and experiments and activities, such as a "short-term memory game," that make learning about the brain fun and interactive. Suggested grade levels are provided for the activities.

**New Mexico Museum of Natural History and Science**
http://www.nmmnh-abq.mus.nm.us/nmmnh/nmmnh.html

This site features information about exhibits along with informative articles by experts. This science center has a fantastic dinosaur collection and its website contains excellent resources, especially about dinosaurs in New Mexico.

**Newton's Apple Page**
http://ericir.syr.edu/Projects/Newton/

The site of the popular television program has twenty-six lessons—from hang gliding to Arctic survival—with information and activities!

**Office of Transportation Technologies KIDS' Page!**
http://www.ott.doe.gov/coloring/kids.html

The U.S. Department of Energy's Office of Transportation Technologies sponsors this page for elementary school students about alternative fuel cars and helping the environment.

**Official Homepage of the Birmingham Zoo**
http://www.bhm.tis.net/zoo/

This site features pictures and facts about animals. Students can even "Ask the Zookeeper!"

**Optics for Kids: Science and Engineering**
http://www.opticalres.com/kidoptx.html

Teachers can send their students here to learn about light, lenses, and more. The site contains a list of optics-related topics and the student chooses what he or she wants to learn about. This site is probably best for students in middle school and above.

**Oregon Museum of Science and Industry**
http://www.omsi.edu

Highlights of this Portland museum's website include an excellent section of teacher resources and a science trivia quiz called "Whatnot."

**The Particle Adventure**
http://pdg.lbl.gov/cpep/adventure.html

An interactive tour of the inner workings of the atom sponsored by the Contemporary Physics Education Project, this site also contains hundreds of pages of information aimed at middle-school-aged students about the particles that make up atoms.

**Quest**
http://quest.arc.nasa.gov/project/overview.html

This children's site from NASA lets users learn about the people in NASA, view live pictures from the Hubble telescope, and much more. Using this site, classes can do projects through the Internet.

**Safari Touch Tank**
http://oberon.educ.sfu.ca/projects/safari/3DTouchTank/3dlib/squirt.html

This site has pictures of sea creatures. Kids can click the mouse on one of the pictures to learn what it is and find links to other sites for information.

**San Francisco ZOO**
http://www.sfzoo.com/

This site includes a virtual tour of the zoo, a children's nature trail, information about zoo programs, and a section devoted to conservation and zoological science.

**Santa Barbara Museum of Natural History**
http://www.sbnature.org/

At this site, users will find an online form to report sightings of marine mammals, information about marine mammal classification, information about exhibits and programs, and exquisite graphics and photographs.

**Science and Nature for Kids**
http://kidscience.miningco.com/

Sponsored by The Mining Company, this Internet gateway for kids features weekly articles and events, a resource list organized by categories, and an annotated "best of the Net guide."

**Science Museum of Minnesota**
http://www.sci.mus.mn.us/

This site contains information and activities based on interactive workshops and exhibits conducted with children at the science center.

**The SciTech Quark Machine**
http://www.town.hall.org/places/SciTech/qmachine/index.html

Here kids can learn about quarks, learn about the different kinds of quarks, and make their own particle.

**The Search for Ice and Snow**
http://www.lhs.berkeley.edu/SII/SII-EarthIce/ski.1.homepage.html

Developed by the Lawrence Hall of Science, this site contains a science activity for kids in grades 4–12, in which they search for places that have ice and snow. Visitors are given a step-by-step guide and links to resources to help them find ice and snow.

**Toys in Space**
http://www.observe.ivv.nasa.gov/nasa/exhibits/toys_space/toyframe.html

This website documents what happened when astronauts took simple children's toys on a shuttle mission to conduct experiments that would help teach children about force and motion.

**The University of Florida Book of Insect Records**
http://gnv.ifas.ufl.edu/~tjw/recbk.htm

This is an online book of insect "champions." Each chapter deals with a different entomological category, for example, the fastest flyer, the longest life span, the most spectacular mating, and least discriminating sucker of vertebrate blood!

**UT Science Bytes**
http://loki.ur.utk.edu/ut2kids/

Science Bytes is designed for elementary and secondary school students and teachers. Each installment describes the work being done by scientists at the University of Tennessee.

**The Virtual Birder**
http://magneto.cybersmith.com/vbirder/

This fun site is an online newsletter for bird lovers. Each month a different topic is highlighted. Users can also take a virtual birdwatching tour in the section called On Location. Kids can select their skill level and image size, zoom in on the birds, and test their knowledge, collecting points for each correct answer on the way.

**VolcanoWorld**
http://volcano.und.nodak.edu/vw.html

The University of North Dakota's VolcanoWorld site provides on-the-scene descriptions, photos, analysis of recent eruptions, an "ask a volcanologist" section, and a kid's page. Users can also choose to have VolcanoWorld e-mail news of the latest eruptions.

### The Weather Dude

http://www.nwlink.com/~wxdude/

Seattle's KSTW-TV weather forecaster Nick Walker hosts this interesting and fun site that features topical weather information, weather songs, and much more.

### WhaleTimes SeaBed

http://www.whaletimes.org/whahmpg.htm

This site includes fun facts, puzzles, and bibliographies for kids. There's also a section in which kids can write stories and submit stories online.

### Wonderful World of Bugs

http://www.ex.ac.uk/~gjlramel/welcome.html

This site is good for both teachers and kids. Teachers can find detailed information on specific insects and a guide to books on entomology, and use links to other sites and to clubs and societies. Kids can get into the site's Bug Club to find out more about insects, view the bug of the week, read the club's newsletter, or join the club.

### Your Name in Hieroglyphics

http://www.iut.univ-paris8.fr/~rosmord/nomhiero.html

Kids who visit this site can see how their names would have been written in hieroglyphics by Ptolemaic Egyptians. It is written in both French and English, so if you have students who know French or are learning it, they might have fun just trying to translate the French text.

## Websites for Librarians and Teachers

Throughout this book we provide Internet resources that relate to the chapter topics. For example, Chapter 1 has a list of sites that will help you learn more about science education reform and standards-based education. Similarly, Chapter 7 provides websites that will give you more information about fund-raising. In this section, you will find resources that you can use to get more information from and about the Internet. After spending even a short amount of time surfing the Internet, you may come away with the paradoxical feeling that there is too much information available on the Internet, and yet it's difficult to find what you need or want. This is particularly true when you are looking for information about the Internet itself. The Internet is a very self-reflective medium. Much like Narcissus, it enjoys looking at itself. If you want graphic proof of this, go to any search engine and search for sites that contain the word *Internet.*

With so much available content, it pays to sharpen your Internet skills so that you can make the most effective use of your online time. To help you get the most out of the Internet, we've compiled a list of online Internet guides. The sites on the list use a variety of styles—some take a scholarly approach while others are more lighthearted. Because we all have different learning preferences—some of us like to chart our own paths while others may prefer a little hand-holding—there should be something here for everyone.

**"Back To School:" The Electronic Library Classroom 101**
http://web.csd.sc.edu/bck2skol/fall/fall.html

This is billed as a free electronic library classroom created by Ellen Chamberlain, Head Librarian, University of South Carolina Beaufort; and Miriam Mitchell, Senior Systems Analyst, USC Columbia. It consists of thirty lessons and several addenda and is very comprehensive. If you are looking for a thoughtful, extended introduction to the Internet, be sure to give this a try.

**Cyberspace Companion: Introduction to the Virtual Classroom**
http://w3.aces.uiuc.edu/AIM/scale/

The Cyberspace Companion is an introduction to using computers in classes at the University of Illinois at Urbana-Champaign. It contains tutorials for popular software programs, such as Adobe Acrobat Reader; various Web browsers, such as Netscape; word-processing programs, such as Microsoft Word and Word Perfect; spreadsheet programs, such as Lotus 1-2-3 and Microsoft Excel; and operating systems, such as Windows, Unix, and DOS. These are online tutorials that offer step-by-step instructions. Although the tutorials are targeted to students at the university, they can be used by anyone. The tutorials are well written and the graphics are excellent.

**The 15 Minute Series**

http://rs.internic.net/nic-support/15min/

This collection of Internet training materials is provided as a service to the research and education communities by the InterNIC and the Library and Information Technology Association (LITA), a division of the American Library Association (ALA). The collection is intended to help members of the research and education communities incorporate and support the growing role of the Internet in their day-to-day operations and activities. Each of the 15-Minute Series modules is structured as a mini–slide presentation and is designed to answer a specific Internet-related question. Librarians who are responsible for training others to use the Internet should find this site useful.

**Finding Information on the Internet: A Tutorial**

http://www.lib.berkeley.edu/TeachingLib/Guides/Internet/FindInfo.html

Hosted by the University of California, Berkeley, this site provides a platform for beginners to the Internet, the World Wide Web, and the Netscape browser. The site recommends and explains effective search strategies applicable to any research interest, and provides descriptive information on other search tools and strategies.

**Finding Your Way on the World Wide Web**

http://www.mwc.edu/ernie/fpcug01.html

Ernest Ackerman, author and computer science faculty member at Mary Washington College, offers this well-written and user-friendly introduction to the Internet. It contains the following sections: (1) What is the World Wide Web? (2) Some Navigational Aids; (3) Directories—Subject Trees; (4) Search Engines; (5) Discussion Groups; and (6) Usenet News. Ackerman's web page also has some other useful Web guides, including a guide to Internet discussion groups. Go here if you need to get up to speed on Net etiquette.

**The ICYouSee Home Page**

http://www.ithaca.edu/Library/Training/ICYouSee.html

Designed as a project of the Ithaca College Library, this site is a self-guided World Wide Web training page. Its primary purpose is to provide an introduction to the World Wide Web. Its secondary purpose is to offer a survey of interesting and useful places that are out there on the

Web. Among the excellent features are: an easy-to-use glossary from which you can select terms from a chart and link to the explanation; a useful guide to different types of websites; a problem-solving guide that explains the most common types of error messages you are likely to encounter; and a quiz to help you assess how much you've learned.

**Internet Basics for Educators**
http://www.voicenet.com/~bertland/compf/iclass.html#top

This site calls itself an on-ramp for educators who are planning to venture forth on the information highway. It's actually a very good place to get on the Web. It includes basic information about the Internet and a special section that helps educators use the Internet with students. It's brought to you by the Virtual Branch of the Stetson Middle School Library in Philadelphia. Don't just stop at the Internet introduction on this site. Check out all the other great resources compiled for Internet users by the library staff at Stetson Middle School.

**Internet Learners Page**
http://www.clark.net/pub/lschank/web/learn.html

Users can find the following sections at this site: Starting Points; Ethical and Legal Issues; Internet Guides; Internet Lists of Lists; HTML (includes Perl, cgi, images, etc.); Searching the Internet (includes e-mail); and World-Wide Web Resources. The resources on this page are not only a good beginning; they can also lead to deeper involvement with the Internet, including constructing your own web pages and learning to use programming languages, such as HTML.

**The Internet Public Library (IPL)**
http://www.ipl.org

The mission of this free service is to find, evaluate, select, organize, describe, and create quality information resources. The site has sections devoted to resources for youth and teens, and includes a special section for kids that points them to resources on a variety of subjects, including science and math. Its reference center includes many original resources created by IPL staff. Also, a Science Fair Resource links to websites that show users how to do a science fair project.

**Internet Resources for Institutional Research**
http://apollo.gmu.edu/~jmilam/air95.html

This page was designed for university researchers, but that doesn't mean it's not well designed and easy to use. Maintained by George Mason University to assist institutional researchers and faculty and students in higher education, this guide is for you if you're looking for a thorough and rigorous introduction to the Internet.

**Internet Tour**
http://www.infolane.com/nm-library/itblcon.html

This great site was designed by the librarians at the Newark Memorial High School Library in California to get their students online. It includes practice exercises and covers just the right amount of information to get you up and running on the Web. In fact, the librarians at Newark Memorial High School require that students complete the tour and pass the online quiz in order to gain access to the Internet at the library. Fortunately for their students, the site is a lot of fun.

**Librarian's Index to the Internet**
http://sunsite.berkeley.edu/InternetIndex/

Make sure you surf this fantastic resource from the University of California's Digital Library SunSITE service. Search or browse this huge subject index of Internet reference materials and you'll find links to several Internet Reference Desks. Topics range from the arts to women and cover both academic and general interest subjects. Highlights of this site include the Internet Information (including graphics) and Internet Searching pages. Make sure to bookmark this one because the latest additions to the index are posted weekly in a New This Week section.

**Peter Milbury's School Librarian Web Pages**
http://wombat.cusd.chico.k-12.ca.us/~pmilbury/lib.html

This site, created by Chico (Calif.) Senior High School's Peter Milbury, contains links to web pages created, maintained, or both by school librarians. It is divided into various categories, such as K–12 schools, personal pages, and curriculum-related sites. It also contains information about creating your own website. After you've done that, you can add your page to the site.

**Teaching Students to Think Critically about Internet Resources**
http://wombat.cusd.chico.k-12.ca.us/~pmilbury/lib.html

This website contains an online tutorial designed for faculty and teaching assistants at the University of Washington. Especially good are the activities that focus on learning to evaluate information on the Internet. Going over the material on this site can help you help students, and it can also help you become more aware of the intricacies of the Internet.

**Understanding and Using the Internet/Beginner's Guide to the Internet**
http://www.pbs.org/uti/

This site, combining two resources from PBS, provides a nice introduction to the Internet through links to useful sights and utilities. Included are access to helper applications, such as Shockwave, Adobe Acrobat, and QuickTime for Windows. You can also read background articles and browse online transcripts of the PBS programs on which these sites were based.

**University of Wisconsin–Eau Claire McIntyre Library: Ten C's for Evaluating Internet Resources**
http://www.uwec.edu/Admin/Library/10cs.html

This is an excellent guide for teachers, students, and librarians alike. This primer on evaluating the content of websites is succinct and to the point. It urges Internet users to employ critical thinking in evaluating the credibility of Internet content. This kind of information is essential for getting the most out of the Internet.

**Where the Wild Things Are: Librarian's Guide to the Best Information on the Net**
http://www.sau.edu/CWIS/Internet/Wild/index.htm

This site is a must for librarians. Among the valuable features are a special list of resources on disabilities, hot topics, full texts of books, graphics and picture resources, and a reference desk for answering questions. Other useful sections include Internet Guides and Resources; Librarian's Gold Mine; and What's New on the Net.

**The WMHT/Troy City School 14 Internet Testbed Project**
http://www.wmht.org/trail/

This Internet guide was designed specifically for K–5 teachers by public broadcaster WMHT for teachers at Troy City School 14, located in the Capital Region of New York State. The authors decided that the best way to introduce K–5 educators to the Web was to show them what other elementary schools are doing on the Internet. Each stop along the CyberTrail includes a description of a website, an explanation of why the site exemplifies effective use of the Internet in K–5 education, and a direct link to the featured site. This one's a real winner.

# 6

# Inquiry-Based Learning in the Library

MARIA SOSA

Libraries and inquiry-based learning are familiar partners. After all, for many of us, the library is where we go to find answers to questions. Sometimes, as in the case of homework assignments, the need to find the answers is driven by external forces. Just as often, however, the questions for which we seek answers are driven by personal curiosity. Librarians are familiar with inquiry in nearly all its manifestations and are not surprised by the questions that are asked or by who asks them. For example, a librarian may know that Lisa is interested in car engines or that Carl reads every book on astronomy that he can get his hands on. In fact, no one but the librarian may know this about Lisa and Carl. This special relationship between a librarian and a child is one reason why we believe that librarians can play a vital role in bringing about meaningful reform in the way children learn science.

Inquiry is central to science learning. When engaging in inquiry, students ask questions; plan and conduct investigations; use appropriate tools and techniques, including reference resources, to gather data; think critically and logically about relationships between evidence and explanations; construct and analyze alternative explanations; and communicate scientific arguments and results. In this type of learning environment, students actively develop their understanding of science by combining their results and scientific knowledge with reasoning and thinking skills.

Activities involving inquiry can vary in duration. Some can be done in twenty to sixty minutes or take several days or weeks. Longer activities can begin in the library and then be used as science-fair projects or take-home activities.

Inquiry-based activities do not have just one answer. The goal is to change the perception that science consists merely of rote memorization that leads to precise, unchanging results. Inquiry-based science activities should promote discovery and encourage cooperative learning by helping students find their own solutions to problems.

This chapter contains some excellent resources that can help you locate activities appropriate for inquiry-based learning. The projects in Chapter 8 can also stimulate ideas about ways to conduct inquiry-based activities in the library. Be sure to conduct a trial run of each activity before you try it with a group of students. If you are unsure of or lack confidence in your own understanding of a scientific concept, conduct an exploration of your own. Ask questions, read books, or consult an expert. Remember how you came to understand the concept as you guide student explorations. This will help you anticipate some of the questions children might have. The following basic principles should help you integrate science into your library programs for young people:

*Integrate science and mathematics into your daily activities.* Make science and mathematics a familiar part of the library. Help children feel that these topics are related to everything that they learn and do by making the library a science- and math-friendly place.

*Provide time for reflection and discussion.* It is very important to provide reflection time for children after they complete an activity. This helps students build their own understanding, an important component of effective science instruction. Asking leading questions, such as "What happened in your investigation?" or "Why do you think this happened?" or "What do you think will happen if we do this?" will help to stimulate discussion. The activities at the end of this chapter contain questions to guide you.

*Encourage students to verbalize their observations and explanations.* Communication is a very important part of science. Use a tape recorder to encourage children to share their observations. In so doing, you place children in an environment in which they can talk with peers and adults as well as listen to language as they express their ideas. Vocabulary and language patterns used in natural situations are nurtured through practice—talking, listening, and responding. The opportunity to discuss feelings and reflect on events enhances children's ability to express themselves in words.

*Have students work in small groups.* It is good practice to form smaller groups—three to four works well. Small groups not only help maintain discipline and facilitate the flow of activities, but also give students the opportunity to interact, share ideas, and work together cooperatively to solve problems.

*Make predictions.* Discuss with students the concepts that will be explored in the activity. Ask students what they already know about the topic. Ask them to describe some things they think might happen.

*Keep a science journal.* Ask students to keep a journal in which they write down their questions, predictions, observations, and findings. These journals can be shared with other children, with science mentors, or with family members and can be the beginning of successful science fair projects. Keep your own journal in which you record your experiences, both as a fellow explorer and as an exploration guide. Even though you have conducted an activity many times, the opportunity is always there to learn something new or to see something in a different light. A journal will help you reflect upon this process.

*Use mathematics in all aspects of scientific inquiry.* Mathematics is essential to asking and answering questions about the natural world. Mathematics can be used to ask questions; to gather, organize, and present data; and to structure convincing explanations.

*Create a display of books related to the topic of exploration.* Be creative—include fiction, nonfiction, activity books, reference books, and poetry. A bibliography of books that support inquiry can be found at the end of this chapter.

*Base future explorations on student-generated questions.* This will not only help promote inquiry-based learning, but also help students interested in science-fair projects use their own curiosity to develop topics.

*Use a variety of instructional materials and resources.* Use children's science books to help children learn about a variety of science topics. Use hands-on activities to promote inquiry and problem solving. Play educational games, including puzzles, and show videos that demonstrate science phenomena or that portray habitats and environments that are not in students' everyday experience. Collect articles from newspapers or popular journals that focus on science, health, or technology. Have the students read the health and science sections of newspapers and magazines, write reports, and share their reports with the group.

## Tips for Guiding the Learning Experience

Leading inquiry-based learning can be scary, because the adult facilitator has to step back and allow the learners to explore at their own pace and in their own way. Here are some tips that will make the activities run more smoothly.

1. Even though students will be free to conduct their own inquiries, a certain amount of order is always needed when working with groups of children. Begin each session after all participants are seated, quiet, and ready to listen to instructions.
2. Introduce each activity. This doesn't mean that you should explain what will happen or tell kids what results they should get. You should, however, set the activity in a context that will be motivating and easy for the children to understand. For younger children, you can connect the activity to a poem or to a work of fiction.
3. Have children work in groups whenever possible. Choose activities that use inexpensive materials so that each group can have everything that it needs and the kids can work at their own pace. This will avoid situations in which children are waiting to use a piece of equipment. During "down" times, kids can get distracted or bored.
4. If an activity has several steps for which different materials are needed, pass out the necessary items at the beginning of each step. This will prevent kids from getting distracted by materials that they will not be using until later.
5. Make sure that everyone has enough space in which to work and that the workstations are clearly established. If there are parts of

the library in which you don't want children to work or which may be too distracting, make sure that the workstations are situated away from these areas.

6. Hand out an activity sheet for the kids to work on. Make sure the sheet has simple instructions and questions to guide the inquiry, but doesn't include answers. If you are adapting an activity from a book and it includes a "why it happens" section, don't give that to the learners. You may wish to pass out the explanation after the activity, but that is certainly not necessary. It may be preferable to ask students to find resources in the library that relate to the activity they have just conducted.
7. Make sure that the children stay busy. Some groups may finish before others, either because they work more quickly or because they work more superficially. When a group appears to be done, ask the children about what they have done and pose questions that lead to further areas that they can explore. As children ask questions, turn the questions back to them by asking for more information or for examples of what they have observed.
8. When discussing findings, make sure that kids understand that communicating the results of their inquiry is a critical part of science exploration. Also, make sure they understand that all the results may not be the same. Use discussions of findings to help frame further explorations. Remind children that scientists often test their colleagues' ideas. Children may want to try things that others have done to see if they get the same results.
9. Provide closure to the activity. Even though the inquiry may be open-ended, kids need to reflect and summarize. This is especially important for younger students.
10. Have sources on hand to answer questions. These can include tradebooks related to the topic or volunteer scientists. Make sure the volunteers understand that in inquiry-based learning, they are supposed to help facilitate student inquiry and not to provide quick answers. "I don't know, but let's see if we can find out," is a very useful response in many cases.

The science standards emphasize that students should be provided opportunities to engage in both full and partial inquiries. In a full inquiry, as noted at the beginning of this chapter, students begin with a question, design an investigation, gather evidence, formulate an answer to the original question, and communicate the investigative process and results. In partial inquiries, however, students develop understandings about selected aspects of the inquiry process. After a teacher-led demonstration on a natural phenomenon, for example, students might be asked to recognize and analyze several alternative explanations for what the teacher has presented. Another example of partial inquiry would be to ask students to describe how they would design an investigation.

Instructional activities should engage students in forming an understanding of the question being investigated. Students should know what the question is asking, what background knowledge is being used to frame the question, and what they will have to do to answer the question. The students' questions should be relevant and meaningful for them. To help focus investigations, students should ask such questions as: "What do we want to find out about?" "How can we make the most accurate observations?" "Is this the best way to answer our questions?" and "If we do this, then what do we expect will happen?"

Students should present the results of their inquiries in oral or written reports. Students' discussions should center on such questions as: "How should we organize the data to present the clearest answer to our question?" or "How should we organize the evidence to present the strongest explanation?" During these discussions, learners shape their ideas and explain their data, tapping into background knowledge where appropriate. This provides students with an opportunity to reflect on how science is done and gain firsthand knowledge of the rules of scientific thinking.

Communication is key to assessing a student's understanding of science content. Students need opportunities to use the knowledge and language of science to communicate scientific explanations and ideas. Writing, labeling drawings, building spreadsheets, and designing computer graphics are all components of science learning that can be developed in a library-based program. As part of this communication process, students should receive constructive feedback on the quality of their thought and expression and the accuracy of their scientific explanations. Students can work with science mentors who can help them refine their communication skills.

At the end of this chapter are three inquiry-based activities that you can use in the library. Each activity has a set of instructions for the leader as well as pages for students to guide them through their inquiry. Use these activities as models to design your own inquiry-based lessons using some of the resources in our bibliography.

## Bibliography

### Books That Support Inquiry

Ardley, Neil. *101 Great Science Experiments.* Blue Ridge Summit, Pa.: TAB, 1995.

———. *The Science Book of Color.* New York: Harcourt, 1991.

Churchill, Richard. *Amazing Science Experiments with Everyday Materials.* New York: Sterling, 1991.

Doherty, Paul, and Don Rathjan. *The Magic Wand and Other Bright Experiments on Light and Color.* Exploratorium Science Snackbook Series. Exploratorium Teacher Institute. New York: Wiley, 1995.

Doherty, Paul, Dan Rathjen, and the Exploratorium Teacher Institute. *The Spinning Blackboard and Other Dynamic Experiments on Force and Motion.* New York: Wiley, 1996. 112pp.

The Franklin Institute Science Museum. *The Ben Franklin Book of Easy and Incredible Experiments.* New York: Wiley, 1995.

Friedhoffer, Bob. *Physics Lab in a Hardware Store.* Danbury, Conn.: Watts, 1996.

———. *Physics Lab in a Housewares Store.* Danbury, Conn.: Watts, 1996.

———. *Physics Lab in the Home.* Danbury, Conn.: Watts, 1997. 80pp.

———. *Toying Around with Science: The Physics behind Toys and Gags.* Danbury, Conn.: Watts, 1995. 95pp.

Gardner, Robert. *Science Projects about Electricity and Magnets.* Springfield, N.J.: Enslow, 1994. 128pp.

Gilbert, Harry, and Diana Gilbert Smith. *Gravity, the Glue of the Universe: History and Activities.* Englewood, Colo.: Teacher Ideas Press, 1997. xv+206pp.

Lauber, Patricia. *What Do You See and How Do You See It? Exploring Light, Color and Vision.* New York: Crown, 1994.

Newton, David E. *Consumer Chemistry Projects for Young Scientists.* New York: Watts, 1991.

Reading Is Fundamental. *Habitat Lab.* Teacher's Guide. Dubuque, Iowa: Kendall/Hunt, 1996.

———. *Water Lab.* Teacher's Guide. Dubuque, Iowa: Kendall/Hunt, 1996.

Taylor, Beverly A. P., James Poth, and Dwight J. Portman. *Teaching Physics with Toys: Activities for Grades K–9.* Blue Ridge Summit, Penn.: TAB, 1995. viii+296pp.

Time-Life Books. *Physical Forces.* Alexandria, Va.: Time-Life, 1995. 151pp.

Tomecek, Steve. *Bouncing and Bending Light: Phantastic Physical Phenomena.* New York: Scientific American Books for Young Readers, 1995. 48pp.

Vecchione, Glen. *Magnet Science.* New York: Sterling, 1995. 128pp.

Wellington, Jerry. *The Super Science Book of Forces.* New York: Thomson Learning, 1994. 32pp.

Wiese, Jim. *Detective Science: 40 Crime-Solving, Case-Breaking, Crook-Catching Activities for Kids.* New York: Wiley, 1996.

Wood, Robert W. *Mechanics FUNdamentals: FUNtastic Science Experiments for Kids.* Illus. by Rick Brown. Science FUNdamentals Series. New York: McGraw-Hill, 1997. xii+148pp.

Zubrowski, Bernie. *Making Waves: Finding Out about Rhythmic Motion.* New York: Morrow Junior, 1994. 96pp.

## Internet Resources

The following Internet sites offer hands-on science activities.

### Access Excellence
http://www.gene.com/ae/

This network for high school biology teachers has materials that can be searched. This page has an activities exchange, provides for online discussions and seminars, and maintains message boards for teachers.

### Bill Nye the Science Guy
http://nyelabs.kcts.org/

Visitors to this site will find a selection of simple activities on topics ranging from life science to physical science to earth science.

**Blue Web'n Learning Applications**
http://www.kn.pacbell.com/wired/bluewebn/

On this page, users can search for a specific educational tool, such as an Internet lesson or activity for a specific level (for example, a physics lesson for grades 7–12).

**CEA Science Education Home Page**
http://www.cea.berkeley.edu/Education/

This site is a gateway to a variety of K–12 science sections, including the Internet-based Classroom Resources for K–12. It provides teacher-developed lesson plans that incorporate the Web and includes programs for families and Spanish-language programs.

**Computer as Learning Partner**
http://www.clp.berkeley.edu/CLP.html

The Computer as Learning Partner (CLP) project, an ongoing educational research effort at the University of California at Berkeley, is dedicated to informing and improving middle-school science instruction. At this site teachers can find curricula for thermodynamics, light, and sound units.

**Exploratorium Science Snacks**
http://www.exploratorium.edu/snacks/snackintro.html

Available at *Exploratorium's ExploraNet* is a section of fun experiments called Science Snacks.

**The Explorer**
http://explorer.scrtec.org/explorer/

This site contains a collection of resources for K–12 math and science education. The resources include instructional software, lab activities, lesson plans, and student-created materials and are available in Adobe Acrobat format. The site was developed by the Great Lakes Collaborative and the University of Kansas UNITE group to involve educators and students in creating and using multimedia resources for active learning.

**Kinetic City Cyber Club**
http://www.kineticcity.com/

At this AAAS site, users can find activities that provide an opportunity to practice inquiry skills in the Kinetic City Radio Show archives. The Kinetic City Home Crew Journal and the postcards found in the Adventure Express Kit also provide activities that complement the science and technology standard.

**Little Shop of Physics**
http://129.82.166.181/default.html

This is an excellent source for online physics experiments. Created the physics department at Colorado State University, it has some fun little experiments, some of which you can do right in front of the computer screen.

**Project Labs**
http://www.rohmhaas.com/company.dir/plabs.dir/

This cooperative program for science teachers, sponsored by the Rohm and Haas Company and Chestnut Hill College in Delaware, includes experiments for elementary school, middle school, and high school students. Available as Adobe Acrobat files.

**Space Educators' Handbook**
http://tommy.jsc.nasa.gov/~woodfill/SPACEED/SEHHTML/seh.html

This excellent site from NASA leads teachers to lesson plans, movies, space comics, math lessons, astronomy information, and fun facts. The math may require at least a middle school level.

## ACTIVITY 1

# Color Burst

## What We Will Explore

What are the color components of various colors of ink? In what order do colors separate? How might a mixture of dyes be separated?

## Curiosity Starter

Fill two colorless plastic or glass cups about 3/4 full of water. To one of these add a few drops of blue food coloring and to the other a few drops of yellow. Mix well. (The colors should be pretty deep for a good effect.) Pour about 1/3 of each colored solution into another empty cup. Mix well. What is observed? The new color (green) seems to be a mixture of blue and yellow. To a fourth cup of water, add a few drops of green food coloring. Can you tell the difference between the color in this fourth container and that in the third? Is the green food coloring a mixture of blue and yellow? How can we find out? Try to guide the explorers to think about how to separate a combination of dyes into individual components in order to figure out what the combination is.

## Who This Exploration Is For

Anyone can have fun with this activity. Younger explorers will enjoy making and displaying the butterfly shape. You can also use this as a springboard for discussion and for reading books about colors and butterflies. Older explorers can delve more deeply into the topic of chromatography.

## Materials Needed

### Per person

- 18-ounce wide-mouth plastic cup filled with 1/2 inch (1 cm) of water
- Melitta coffee filters (#6 size) or similar brand and size
- black water-soluble marker (nonpermanent markers for overheads work best)
- scissors
- pencil
- 2 or 3 paper towels
- coleus leaf or other brightly colored leaf (optional)
- other colors of markers, food colorings, etc.

## Safety Considerations

Normal care should be taken when using scissors. Younger explorers should probably use blunt scissors. If you are doing this activity in the library, you may wish to consider the "safety" of library materials when children are using water and markers. Even though little water is used, plan to have a lot of newspaper and paper towels for spills.

## Adaptations for Explorers with Disabilities

- Participants with hearing impairments should be able to do this activity without any modifications other than those necessary for communicating the instructions.
- Children with limited manual dexterity may work with a partner. You can also tape the bottom of the cup to a table surface so that it will be more stable.

## What to Do

- With a pencil, explorers should draw dotted lines on the coffee filter as shown (in order to make a butterfly shape when the paper is later unfolded).

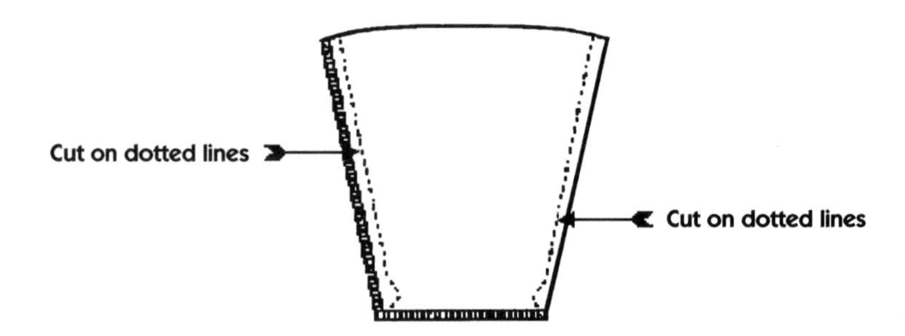


- Cut the filter on the dotted lines. The remaining filter should still be in one piece.
- Using the black marker, explorers should decorate both sides of the filter with a few dots, lines, or other markings. The simpler the pattern, the more striking the results will be. Explorers should be careful not to mark the ribbed bottom edge.
- Have explorers place their filters in the cup of water as shown in the picture. Only the ribbed edge should be in the water.
- Allow the filter to sit undisturbed. Every few minutes, explorers should observe what is happening and make drawings of what they see. Is there a color separation? Is it the same for each mark made?

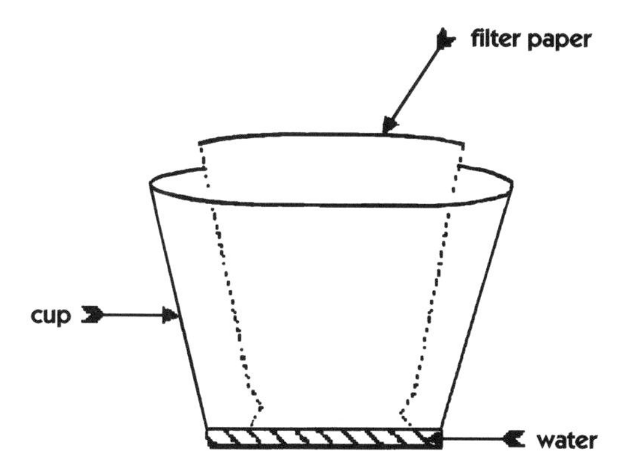


- As soon as the water level has risen to the top of the paper, remove the filter from the cup, and gently open the filter. Compare the filters. Have explorers answer the questions on the activity sheet.
- After you have finished discussing the results, set the filters aside to dry.

## Guiding the Exploration

Even though it takes ten to fifteen minutes for the colors to fully separate, explorers should be attentive to the separation process as it goes on. You may decide to have students record what they observe in five-minute intervals, or you may simply ask guiding questions, such as the following, to keep students focused on the observation.

- What is happening to the black ink? Describe what you see.
- What colors do you see on the filter before you open it up? After you open it up?
- What happened that you didn't expect or that was different from what you expected?

## For Younger Explorers

Younger explorers may lack the attention span to observe the entire color separation process. You may wish to occupy them with a discussion of color or with books about color and discuss the color separation after the water has climbed the paper, using questions such as the following. However, you should have them check the cups occasionally to see how the filter changes over time.

- What colors do you see when you open the filter?

- What shape is the coffee filter when you open it up? Why do you think the activity is called "color burst"?
- What do you think happened to the water? What happened to the black ink?

## Where to Go from Here

- Do you think the same thing would happen if you used red ink? Green ink? Purple ink? Try it.
- What is the effect of temperature on color separation? How can you find out? Try it.
- Does the shape of the filter make a difference in what you observe? Does the kind of mark (dot, line, etc.) make a difference? How can you find out? Try it.
- What if you used a different liquid in the cup? Do you think the colors would still separate? Try it.
- What do you think would happen if you ground up a coleus or other brightly colored leaf, placed a dot of the leaf's pigment on a filter, and repeated the activity? Try it.

## Why It Happens

This activity uses a technique called paper chromatography. The water is absorbed by the coffee filter and rises up the filter. When the water reaches a spot of black ink, it carries the components of the spot up the filter. As the water continues to rise, the components will rise also, but at different rates. Some are more soluble in water than others, and the more soluble ones travel faster. After a time, the various components are at different distances from the original spot. They have now been separated from one another. Substances that can cause other substances to dissolve, like water can, are called solvents.

Water is the simplest solvent to use in paper chromatography. Not all components in a sample may dissolve in water. Alcohol and ammonia are also solvents, but should not be used with young children. Scientists use chromatography frequently to separate and identify the component parts of solutions. It is a valuable tool for helping us understand what makes up the many solutions in our environment.

**NOTES**

# Color Burst

## You Will Need

- 18-ounce wide-mouth plastic cup filled with 1/2 inch (1 cm) of water
- Melitta coffee filters (#6 size) or similar brand and size
- black water-soluble marker (nonpermanent markers for overheads work best)
- scissors
- pencil
- 2 or 3 paper towels
- your science journal

## What to Do

- With a pencil, draw dotted lines on the coffee filter, as shown in the picture.
- Cut the filter on the dotted lines. Your filter should still be in one piece.

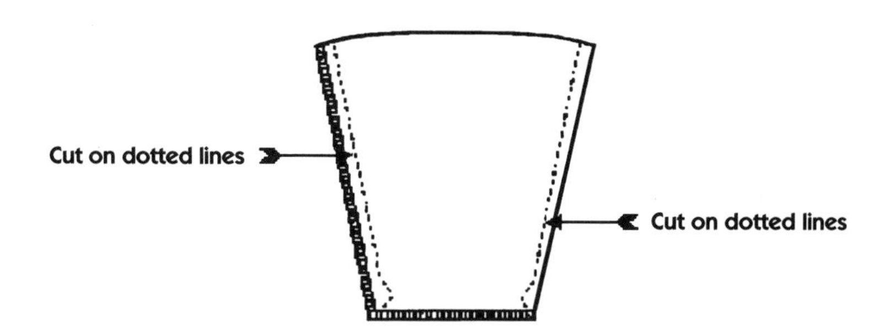


- Using the black marker, decorate both sides of the filter with a very simple design of dots, lines, or other markings. Be sure not to mark the ribbed bottom edge.
- Place your filter in the cup of water as shown in the picture. Only the ribbed edge should touch the water.
- Allow the filter to sit undisturbed. Check at regular intervals to see what is happening and record your observations in your science journal.
- Once the water level has risen to the top of the paper, remove the filter from the cup, and gently open the filter. What do you see? Answer the following questions in your science journal.

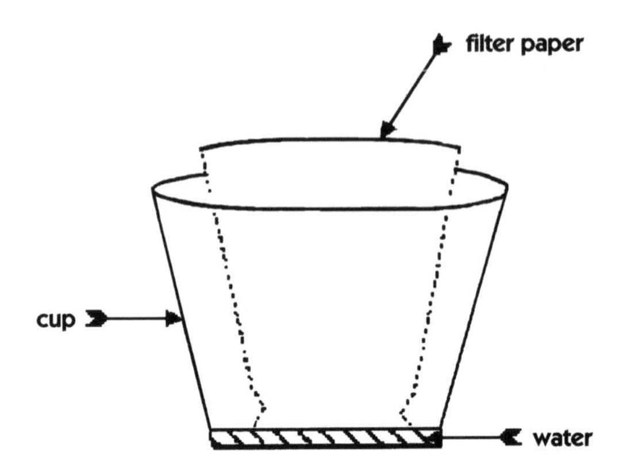


1. What happened to the black ink? Where is it on the filter paper?
2. What colors do you see on the filter after you open it up?
3. What shape is the coffee filter when you open it up?
4. What happened in this activity that you didn't expect or that was different from what you expected?
5. Is the shape of the coffee filter important? What would happen if the shape were different? Try it.
6. Do you think the same thing would happen if you used red ink? Green ink? Purple ink? Try it.
7. Will other colored liquids act the same as the inks? How about food coloring? Other kinds of inks?
8. What would happen if you repeated the activity using the pigment of a brightly colored leaf, such as a coleus. How would you get the pigment on the coffee filter? Try it.
9. Does this activity suggest any other questions to you? What are they? How do you think you can find the answers?

## Communicating Science

In your science journal, write a report on what you learned in this activity and on your further explorations. Share your results with the rest of the class. How are their results the same as yours? How are they different?

**NOTES**

ACTIVITY 2

# Floating Eggs

## What We Will Explore

Do the properties of water change when a substance is dissolved in it? In particular, since some objects float on water and others sink, how will floating and sinking behaviors be changed, if at all, by dissolved substances in the water?

## Curiosity Starter

Demonstrate that dissolving a solid in a liquid doesn't always cause the volume of the liquid to increase. Put a long, thin stirrer (such as a cake tester or a kabob skewer) into a clear plastic or glass drinking glass. Fill the glass all the way to the top with water. Measure about two tablespoons of kosher salt into a second empty glass. (We suggest using kosher salt instead of regular salt, because kosher salt dissolves to give a clear solution. Regular salt contains drying agents that make solutions a tiny bit cloudy.) Ask your explorers to decide whether they think you can put all of this salt into the glass of water without making it overflow.

Gently sprinkle a little of the salt onto the surface of the water. Using the stirrer, gently stir the water until the salt dissolves. Stir very carefully so that the water doesn't splash out of the glass. Continue adding the premeasured salt to the full glass of water, stirring gently after each addition until the salt dissolves. How much of the salt can you dissolve without overflowing the glass? (If you are careful not to splash any of the solution, you will probably be able to dissolve all the salt into the water.)

Since there has been no change in the volume of the liquid (that you could detect), how is the solution different from the water without the salt? Try to get the explorers to suggest that the mass (weight) of the solution must be larger, since all the solid is now in the glass, as well as all the water you began with. What might be different about the properties of the solution compared to the water with which you started? What might be tested? See if your explorers know that objects (including people) float better in salt water (the ocean) than in fresh water. Try to get them to suggest a floating/sinking exploration with various dissolved solids (salt and sugar are good to begin with) and some interesting test object, such as an egg.

## Materials Needed

### Per group

- a tall, clear plastic or glass drinking glass or similar container (One-liter clear plastic soda bottles with the tops cut off and the labels removed are excellent.)

- a cup of salt (kosher salt works best) or a cup of sugar
- an uncooked, unbroken egg (in the shell)
- spoons for measuring and stirring
- paper or cloth towels for cleanup
- pitcher of water or sink for filling containers
- sink for discarding solutions and plain water. If this activity is done outdoors, the water can be poured onto gravel or plain dirt (dilute salt water first). Do not pour salt water onto grass or concrete.
- hard-boiled egg (optional)

## Safety Considerations

Spills can create wet floors; be careful of slipping. Don't use glass containers unless you have to because they are breakable and broken glass is a hazard. Make sure your explorers know not to drink any of the solutions they make. It is never a good idea to taste things in the "laboratory."

## Adaptations for Explorers with Disabilities

- Explorers with hearing impairments should be able to do this activity without any modifications other than those necessary for communicating instructions.
- Explorers with mobility impairments can work with a partner.
- Explorers with visual impairments should be able to feel where the eggs come to rest with some assistance from a partner.

## What to Do

### Materials Preparation

- Set up the containers, water pitcher (if necessary), salt or sugar, spoons, eggs, and towels at a station for each team.
- Find a suitable place to empty the water and solutions when the activity is complete.

### To Prepare for Next Time

- Empty, rinse, and dry all glasses, pitchers, and spoons.
- If the eggs aren't damaged, they can still be used for cooking and eating.

## Guiding the Exploration

(See also the questions on the activity sheet.)

As long as some groups choose to investigate sugar and others salt as the dissolved substance, let each group decide which it wants to do. When the explorations are complete, have all the groups share their results. Have the "salt groups" compare their results with the other "salt groups" (and the "sugar groups" with the other "sugar groups") to see how reproducible the results are. Then they can compare the salt and sugar results to see how different or similar they are. Encourage discussion of the reasons for variability (different amounts of water to start with, for example) and for differences between solutes (different densities for the same amount of different solutes). Some questions you might ask during the exploration are:

- What happens to an egg placed in plain water?
- What happens to the salt or sugar as you mix it into the water?
- What happens to an egg placed in salt or sugar water? Does it matter how much salt or sugar is added?
- Would you get the same results with a hard-boiled egg? Try it. What other objects might you try?

## Where to Go from Here

For more activities on water and surface tension, check out the following books: 175 Science Experiments to Amuse and Amaze Your Friends (New York: Random, 1988), by Brenda Walpole; Science Fun with Toy Boats and Planes (Morristown, N.J.: Simon & Schuster, 1987), by Rose Wyler; The Whole Cosmos Catalog of Science Activities, 2nd ed. (Reading, Mass.: Addison-Wesley, 1990), by Joe Abruscato and Jack Hassard; and Bet You Can: Science Possibilities to Fool You (New York: Avon, 1983) and Bet You Can't: Science Impossibilities to Fool You (New York: Avon, 1983), two books by Vicki Cobb and Kathy Darling.

## Why It Happens

In the Curiosity Starter, you should have found that quite a large amount of salt could be dissolved into an already full glass of water without making the water spill out of the glass. When salt, or many other substances, dissolves in water, the tiny particles (ions) of salt find places to fit in between the water molecules. Therefore, the water level doesn't have to rise in order to hold the salt. If you keep adding salt, eventually you will reach a point when no more salt will dissolve, and at that point you will have a saturated solution. After that, if more salt, or anything else, is added to the water, the water level will have to rise to make more room for it.

An egg sinks in plain water because it is denser than water. The egg weighs more than the water it displaces. (Note: An egg that floats in plain water is not a fresh egg! Use another egg and discard the first one.) The egg floats in salt or sugar water, if it is salty or sugary enough, because the egg is less dense than the salt or sugar water. The egg weighs less than the solution it displaces and, therefore, there is a large enough buoyant force to hold it up. By adding salt or sugar to the plain water, you increased the amount of material taking up the same amount of space, so the density of the liquid has increased.

## NOTES

# Floating Eggs

## You Will Need

- a tall, clear plastic container
- a cup of salt or a cup of sugar
- an uncooked, unbroken egg (in the shell)
- spoons
- paper or cloth towels for cleanup
- pitcher of water or sink for filling glasses

## What to Do

To start, add water to your container to a depth of 4 to 6 inches (8 to 12 cm). Using a spoon, lower an uncracked, uncooked egg into the water. Record what you observe in your science journal.

Work with your group to design an experiment to answer these questions:

- Does adding a solute (salt or sugar) to the water change the floating or sinking behavior of the egg?
- If so, what is the least amount of solute needed?

Decide how you're going to run your experiment, and then write a description of your experiment in your science journal. Do the experiment as you have described it, and keep a careful record (perhaps in a data table) of what you do and observe. (Hint: To avoid problems with broken eggs, remove the egg from the water or solution before you add a solute or stir to dissolve it.)

When you have finished your experiment and have answers to the questions (and perhaps more questions of your own), get together with the other groups to exchange information so you can learn from their results and they from yours. Are the results the same? If not, what causes the differences?

## Communicating Science

Write and illustrate a brief article for the library or classroom bulletin board that explains what you did, gives your answers to the questions, and supports the answers with evidence from your journal.

ACTIVITY 3

# Static Electricity

## What We Will Explore

What happens on a dry day when you scuff across a rug and then reach out to touch a metal doorknob (or other metal object)? What happens when you place an inflated balloon against a wall? What happens if you rub the balloon briskly with a piece of carpeting or wool before you place it against the wall? Are these various observations related? What else can you do to explore these phenomena?

## Curiosity Starter

Suggestions for beginning an exploration of static electricity are given in the Guiding the Exploration section. Briefly, begin with a demonstration of a static electricity phenomenon (for ideas, see the activities listed below) followed by a discussion of other phenomena the explorers have experienced or heard about. Find out what questions they have about static electricity phenomena. Some of these may make good jumping off points to one or more of the suggested activities.

## Who This Exploration Is For

Explorers from kindergarten on up should find various parts of these activities and their results interesting. It's more fun to explore together in pairs or small groups, so there will be others with whom to discuss the results and decide on the next thing to try. Depending on how the activity is structured, it can take about fifteen minutes for one activity to an afternoon or more to explore questions generated by the group.

## Materials Needed

### Per group

Many of the same materials are used in several different activities. Also, the activities can be varied by mixing and matching materials. For example, the ping-pong ball (C), tiny pieces of paper (D), and lightweight pieces of cereal (G) all play the same role in their respective activities. Other materials you might try are air-popped popcorn and styrofoam (packing material). Based on your own everyday experiences with static electricity, you can probably think of many other materials that can be substituted for those suggested here.

### A. Stuck-up Balloon

- inflated balloon
- piece of fur, wool, or clean hair from someone's head
- blank space on a nearby wall

### B. Dancing Balloons

- 2 inflated balloons
- 2 lengths of thread or lightweight string a meter or so long (exact length is not critical)
- fur, wool, or hair, as in activity A
- tape
- ping-pong balls, styrofoam, air-popped popcorn, loosely wadded aluminum foil, etc. (optional)

### C. Ping-Pong Ball Pet

- comb or inflated balloon
- fur, wool, or hair, as in activity A
- ping-pong ball
- smooth clear area on a table or the floor
- length of thread or lightweight string a meter or so long (exact length is not critical)
- tape
- styrofoam, air-popped popcorn, loosely wadded aluminum foil, etc. (optional)

### D. Dancing Paper

- comb or inflated balloon
- fur, wool, or hair, as in activity A
- pieces of paper about the size of a small fingernail (Enough pieces can be obtained by tearing up a small piece of notepaper about 5 to 6 cm square.)

### E. Far-out Hair

- comb or brush
- someone's hair (clean, dry, and 5 cm or more in length)

### F. Flying Newspaper

- strip of newspaper about 3 cm wide and 75 to 100 cm long

## G. Snap, Crackle, and Hop!

- clear plastic box about 3 to 5 cm deep (a food storage box will work)
- sheet of aluminum foil larger than the opening of the box
- dry puffed rice cereal (enough to make a layer one-piece deep that covers about half the bottom of the box)
- styrofoam, air-popped popcorn, loosely wadded aluminum foil, etc. (optional)

# Safety Considerations

These vary depending on which activities are done. The activities themselves are harmless. The amount of static electricity that can be generated by rubbing materials together like this is not nearly enough to give anyone a harmful shock. In general:

- Be careful not to frighten anyone by accidentally breaking balloons.
- Be careful not to let very small explorers put objects in their mouths.
- Watch out for paper cuts in activity F.

# Adaptations for Explorers with Disabilities

- Explorers with hearing impairments should be able to do these activities without any modifications other than those necessary for communicating the instructions.
- Explorers with visual impairments may be able to feel some of the effects produced with their hands, especially in activities A, B, C, E, and F. In some cases, touching the objects may cause the electricity in them to be discharged and the effect will be lost.

# What to Do

## Materials Preparation

- Gather the necessary materials for the activities you have chosen to do.
- Arrange a place to do the activities where the explorers will have enough room.
- Make up copies of the student activity sheet.

## To Prepare for Next Time

- Except for inflated balloons, which will not stay inflated, all the materials can be stored in bags or small containers and reused.

## Guiding the Exploration

(See also the questions on the activity sheet.)

The electrical nature of the phenomena your explorers will observe is not obvious. Only with very rare exceptions will anyone see an electric spark generated by any of the activities. You should attempt to make the connection by drawing attention to the first question on the activity sheet—getting a shock when touching a metal object after scuffing across a rug. Leading questions might be:

- Have you ever gotten a shock after walking across a carpet or pulling off a sweater?
- When combing or brushing your dry hair, have you ever found strands of hair following your comb or brush (instead of staying where you want it)?
- Have you ever gotten a shock when you touched the door handle after getting out of a car that has cloth seats?
- What other experiences with static electricity have you had? (This gives a name to the basis for the above phenomena. Names don't necessarily explain anything, but they make it easier to communicate with one another.)

If your exploration is carried out on a very dry day (usually only winter days are dry enough to condense most of the moisture out of the air), you can have your explorers actually do such an activity. If you hold a metal object like a key tightly in your hand, scuff across a rug, and then touch the key to another metal object, you will not feel the shock, but you will be able to hear the crackle and see the spark (best in a darkened room). Thus, the shock (felt if you aren't holding the key), the spark (like a lightning bolt), and the crackle (like the sound of thunder after a lightning bolt) are all evidence that you are dealing with electrical phenomena.

Based on such preliminary activities or previous experience, your explorers should be encouraged to suppose (hypothesize without being told directly) that rubbing an object (like a balloon) with some soft fabric (like a piece of wool or rug) will also produce some electrical phenomenon that they can investigate. The suggested activities are designed to lead explorers to learn that charged objects attract uncharged objects and that like-charged objects repel one another. Perhaps they can think up and carry out activities to try to show that objects of unlike charge attract one another. Set them free to explore the various activities, while you encourage exploration and reinforce good cooperative work and group interaction.

## Where to Go from Here

The book Safe and Simple Electrical Experiments by R. F. Graf contains an entire chapter with thirty-eight activities on static electricity. Some of them are similar to the ones here, but there are many others that you or your explorers might like to try.

## Why It Happens

The phenomena associated with static electricity provide a safe and fun way to explore the electrical nature of matter. In today's model of matter, substances are composed of very tiny atoms. Each atom contains an even tinier positively charged nucleus, in which most of the atom's mass is concentrated. Surrounding the nucleus, in a neutral atom, are just enough very light, negatively charged electrons to balance the charge on the nucleus. The light electrons are moving very fast and are whizzing about the nucleus, forming a sort of "cloud" (like a swarm of bees about a hive) that makes up most of the space occupied by the atom. Each atom is held together by electrical forces. Positive and negative charges attract one another. Electrical forces also hold atoms together with other atoms to form molecules, the tiny particles that make up all the physical substances in the universe (including us).

Almost all the space taken up by any substance is occupied by the fast-moving electrons. When we rub two different substances together, it is likely that one of the substances will hold its electrons a little more loosely than the other will. In that case, some of the electrons can get rubbed off and become attached to the second substance. The first substance will now have fewer negative charges (electrons) than it originally had and will be positively charged (an excess of nuclear charge). The second substance will now have more electrons than it originally had and will be negatively charged. Over time, the first substance will gain electrons from other substances (for example, air molecules) and the second will lose electrons to other substances, and both will once again become electrically neutral. Water vapor in the air is very effective for taking excess electrons from a negatively charged object. That's why activities with static electricity work best on a dry day.

In most substances, electrons can't travel very far. (Metals are the exception; electrons can travel easily through metals. Metals are good conductors of electricity.) Therefore, when electrons are rubbed off onto a surface, they are pretty much stuck on that surface. That is why these activities are said to involve static electricity (electricity that is, in some sense, "standing still").

In activity A, the balloon will probably stick to the wall after being charged by rubbing. When the balloon is rubbed with wool, electrons are rubbed off the wool onto the balloon, so the wool is left with a positive charge and the balloon with a negative charge. As the negatively charged balloon is brought up to the wall, the electrons in the wall are repelled and they tend to move back a little bit. The positively charged nuclei aren't so easily moved and, moreover, are attracted by the negatively charged balloon. As a result, the wall gets slightly more positive at its surface. The positive wall and the negative balloon are attracted to one another, and the balloon sticks to the wall. Because of the charge it acquired when it was rubbed, the balloon can stick to a wide variety of objects (even a person). The time it takes for the charge to be lost can be determined by timing how long the balloon stays stuck to an object. Do all objects act the same?

In activity B, when only one of the balloons is charged, it attracts the other balloon (just as above) and they should swing together and touch. It is likely that some of the electrons will get transferred from the charged balloon to the one that was initially uncharged. Now both of them have a negative charge and they should then repel one another ("like" charges repel). They should swing apart and are likely to end up trying to be farther apart than they were before either was charged. In order to do this, they have to swing so their strings are no longer vertical. But gravity is trying to pull each of them back down. Any small imbalance among these forces or a draft in the room will make the situation unstable and result in the balloons "dancing" around to stay away from one another. When both balloons are negatively charged by rubbing, they repel one another immediately and try never to touch. Other very lightweight objects, such as ping-pong balls, styrofoam, and popcorn, although themselves difficult to rub, can pick up a charge from a charged object like a balloon and then behave similarly to balloons.

In activities C and D, you again have a situation like activity A (or activity B with only one balloon charged). The same sort of separation of charges occurs in the ping-pong ball (activity C) or pieces of paper (activity D) when a charged balloon or comb is brought near. Because these lightweight objects move easily and their positive side is attracted to the charged object, they move toward it. The ping-pong ball can be "pulled" about (very much like leading a pet on a leash) by keeping the charged object just ahead of it as it rolls along. The pieces of paper will probably jump to the charged object and stick momentarily. Some electrons are likely to be transferred to the pieces of paper to give them a negative charge and then they will jump off the negatively charged object. The same thing will happen if you let the ping-pong ball touch the charged object. Now you can move it around by "pushing" it away as it is repelled by an object with a charge of the same sign.

Activities E and F are similar to activity B, in which both balloons have the same charge and repel one another.

## NOTES

# Static Electricity

Have you ever scuffed across a rug and then gotten an electric shock when you reached to touch a metal object like a doorknob? If so, then you have experienced static electricity. There are many easy and fun ways to demonstrate some of the properties of static electricity that can be done with simple materials. Here are some things to try. After you do some of these, see if you can think of others. Remember to record everything you do and observe in your science journal. Also, write down questions that you may have as you work.

In activities that call for rubbing with wool, you can use a sweater, sock, scarf, rug, or anything else that is made out of wool.

## A. Stuck-up Balloon

Blow up a balloon and tie the end so that the balloon stays inflated. Without doing anything else, hold the balloon against the wall and let go. What happens? Briskly rub the balloon across a piece of fur or wool or on your hair (works best if your hair is clean and dry). Without doing anything else, hold the balloon against the wall and let go. What happens? Can you explain what happens? If you assume that rubbing the balloon gives it an electric charge, does that help you explain what happens? How long will the balloon stay stuck to the wall? Will a rubbed balloon stick to other materials and objects besides the wall? Which ones? Which ones will it not stick to? Can you explain why?

## B. Dancing Balloons

Blow up two balloons and tie each one closed. Tie a long thread or string onto the knot of each balloon. Use tape to hang the balloons by their strings from some high support like a doorway. Hang them so they are at the same level and about 5 cm apart. Briskly rub one of the balloons with fur, wool, or your hair as in activity A. Let the balloon go. What happens? Try it again to make sure the same thing happens every time. Can you explain what happens? Is what happens in this activity the same as what happens in activity A? Are there differences between what happens in this activity and activity A? Can you explain the similarities or differences?

Briskly rub both balloons with fur, wool, or your hair. Let them go. What happens? Try it again to make sure the same thing happens every time. Can you explain what happens? Is what happens when both balloons are rubbed the same as what happens when only one is rubbed? Are there differences between the two cases? Can you explain the similarities or differences? What, if any, are the connections between what happens in this case and in activity A?

What happens if you replace the balloons with ping-pong balls? Try it. Briskly rub a balloon with fur, wool, or your hair. Touch it to one of the ping-pong balls and see what happens to the balls. Try it again to make sure the same thing happens every time. Can you explain what happens? What happens if you touch both ping-pong balls with the balloon that has been rubbed? Are there differences between the two cases? Can you explain the similarities or differences? What happens if you touch the ping-pong balls with the fur (after rubbing) instead of the balloon? What happens if you replace one or both of the ping-pong balls with a kernel of air-popped popcorn or a small piece of styrofoam (packing material) or loosely wadded aluminum foil?

## C. Ping-Pong Ball Pet

Place a ping-pong ball on a level surface, such as a tabletop or a smooth, bare floor. Briskly rub a comb or balloon as in activities A and B. Bring the comb or balloon near the ball. What happens? Try it again to make sure the same thing happens every time. Can you explain what happens? What, if any, are the connections between what happens in this case and in activities A and B? What happens if the ping-pong ball is suspended by a long thread? What else could you use besides a ping-pong ball? Are the observations the same? Why or why not?

## D. Dancing Paper

Tear a sheet of paper into small pieces about the size of the fingernail on your little finger. Place the pieces of paper on a table. Briskly rub a comb or balloon as in activities A, B, and C. Bring the comb or balloon near the pieces of paper. What happens? Try it again to make sure the same thing happens every time. Can you explain what happens? What, if any, are the connections between what happens in this case and in the other activities you have done with static electricity?

## E. Far-out Hair

Run a comb or brush through your hair on a cold, dry winter day. What happens to your hair when you hold the comb or brush near it? Can you explain why? What, if any, are the connections between what happens in this case and in the other activities you have done with static electricity?

## F. Flying Newspaper

Starting at the fold, tear across the bottom edge of a full sheet of newspaper so that you'll have a strip about 3 cm wide and 75 to 100 cm long (or whatever the width of the newspaper is). Hang the newspaper strip over one finger at the fold, with the two ends dangling freely. Quickly pull the newspaper strip up between two fingers of the other hand. Watch the dangling strips of newspaper. What do they do? Try it again to make sure the same thing happens every time. Can you explain what happens? What, if any, are the connections between what happens in this case and in the other activities you have done with static electricity?

## G. Snap, Crackle, and Hop!

Place a thin layer of dry puffed rice breakfast cereal on a sheet of aluminum foil. Then put a clear plastic container (about 3 to 5 cm deep) upside down over the cereal. Vigorously rub the upper outside surface of the container (the bottom of the container, because it's upside down) with a piece of wool or nylon. What happens to the cereal underneath the container? Try it again to make sure the same thing happens every time. Can you explain what happens? What, if any, are the connections between what happens in this case and in the other activities you have done with static electricity? What happens if you replace the puffed rice with air-popped popcorn or with small bits of paper? What happens if you don't use the aluminum foil? What else could you try?

## Communicating Science

Review the parts of your science journal in which you wrote about these activities. Try to figure out whether there is an explanation for your observations that seems to fit all of them. Compare your observations and explanations with those of other explorers and see if you can come to an agreement on the explanation. What questions do you still have? What experiments can you think of to try to answer these questions? Where might you search for the answers others have given?

**NOTES**

# 7

# Fund-Raising for Science Activities in the Library

MARIA SOSA

School personnel, particularly school media specialists, often find their programs need resources and materials that don't fit the budget. With the advent of such technologies as CD-ROM, video, and telecommunications, it has become increasingly expensive to maintain a well-stocked, up-to-date media center.

Ideally, every school district would have enough funding to give all departments and programs every desirable resource. Reality, however, is far from ideal. This leaves a media specialist with few options. One option is to operate according to a very strict budget, selecting only the most necessary resources to augment the media collection and tracking down free resources whenever possible. Most media specialists become adept at this. There is, however, another option. School media specialists can raise funds on their own. There are many resources and sources of funding that would be happy to support libraries and media specialists and their projects. If you have a good project idea and the endorsement of your school's administration, you may be able to win grant money!

## Suggestions for Successful Fund-Raising

The first hard and fast rule of fund-raising is: People give money to people. Try to adopt a positive, proactive attitude when planning your fund-raising strategy. If you believe your project should be supported and that it is within your power to obtain support, you will be better able to persuade others of that fact. The following tips can help make your fund-raising efforts pay off.

1. Make your case for support a compelling one. In letters, calls, and visits, emphasize in tangible terms why your activity is important to the local area and how it serves a critical need. Do *not* describe how it works. Focus on aspects of your situation that make it unique.
2. Learn as much as you can about your prospective funder. Understand the funder's goals and objectives and demonstrate how support of your project will help to meet those goals.
3. Anticipate and overcome objections before they arise.
4. Identify how the funder's support will be recognized. This is particularly important with corporate support.
5. Be able to answer questions about how your project will support itself over the long term.
6. When making calls, keep the "pitch" short, but don't forget to ask for the money.
7. Identify opportunities for nonmonetary or in-kind support from corporations and others.
8. Follow-up is critical. Letters should indicate that you will contact recipients within a certain time frame. Be sure that it happens. After a call, send additional information or other materials that might help to increase the prospective funder's level of interest. If a funder indicates "*not* now, but later," be sure to follow up at the appropriate time. If the prospective funder turns you down, find out why. Where did you "miss the boat"?

## Elements of a Successful Fund-Raising Plan

It is important to take a detail-oriented approach to seeking funding. Be as specific as you can when you are planning your project and writing your grant proposal; when funders are awarding grants, they want assurance that a comprehensive plan is in place and is directed by competent individuals. Include the following elements in your project planning.

*Program goals:* Write these clearly and concisely so that a prospective funder can easily ascertain your objectives.

*Financial goals:* Determine "bottom-line" figures for the funding campaign and identify opportunities for in-kind support.

*Prospects and approach strategies:* Amass detailed information on all prospects, and devise a specific plan for approaching each one.

*Board members and other volunteers:* Brainstorm ways of using these individuals on the campaign's development or fund-raising committee.

*Follow-up plan:* Make personal contact either by phone or face-to-face with every funder you approach. Make a concerted effort to identify what worked and what didn't. Modify your plan of approach accordingly.

*Acknowledgment and recognition:* Determine the sort of recognition the funders can expect. Determine opportunities for future partnerships with them.

## Submitting a Successful Grant Proposal

If you have never written a grant proposal before, the prospect can seem daunting. But lots of people write grants, and many of those people are financially rewarded for their efforts. Even if your first grant proposal doesn't get funded, the grant-writing process is a valuable learning experience. There will always be other grants to apply for, which means other opportunities for you to amend your proposal's weaknesses. The following suggestions may help you improve your ability to identify the key elements of a successful grant proposal.

1. Read the proposal information several times and then outline which proposal elements *must* be included.
2. Include a table of contents of proposal elements; refer the reader to the page or paragraph that corresponds to the proposal elements.

3. Have an idea with a new twist. Many people will be submitting grant proposals, so if your idea is not fresh and does not suggest something that will spark interest, your proposal may not be selected.
4. Talk to teachers and administrators about your idea and enlist a support group of readers for your grant proposal. You will need *several* critical readers to edit for you.
5. Check your spelling, grammar, and punctuation.
6. Use a catchy title for your proposal.
7. Show how this proposal can be replicated and generalized to your entire school, county, and state.
8. Include a needs statement; this shows how your project will fill a need in your library or media center.
9. Decide whether you will share materials with others after the project is completed.
10. Indicate how many students will be served by your proposal. The more, the better.
11. If you or your students will develop a product based on the project, describe it carefully.
12. State your budget in numbers and then support it with a clearly written explanation that should include people, stores, and vendors. List any free or donated materials or matching funds that will reduce your budget.
13. Assume the reader knows nothing about what you are describing. Never abbreviate or use acronyms without explaining their meaning. Never use colloquialisms.
14. Be brief and to the point. Readers do not want your life history.

## Types of Support

Countless foundations or branches of corporations exist whose main purpose is to engage in philanthropic activity. They have a certain amount of money each year to use for supporting grant proposals. These philanthropic organizations can be grouped into several categories, some of which will be more helpful to you than others, depending on your project and your own organization. The following sections describe the foundation categories and their characteristics.

## Corporate Foundations

Generally, corporate grants are small, averaging between $5,000 and $10,000. A grant of $25,000 is considered most generous. Initial grants or contributions may be small. You will be judged on your stewardship of any gift, regardless of size.

New groups and projects compete for a limited pool of discretionary funds.

It is often best to solicit in late summer or early fall to coincide with corporate budgeting cycles.

To be successful, projects must clearly relate to corporate bottom-line interests and priorities.

Internal contacts are the key to success; "cold" approaches are most often unsuccessful.

At the national level, expect a minimum six-month turnaround in response to your request. A local plant or division of a corporation may move more quickly.

Historically, corporate funders have awarded grants to educational projects, federated charities (United Way, for example), minority enterprises, and local social services that demonstrate a benefit to the company's employees as well as to the community at large.

## National Foundations

National foundations, such as the Ford Foundation and the Lilly Endowment, are not limited to a particular geographic area in their support.

National foundations are usually quite large, with assets of $25 million or more.

Programs or proposals with national or at least regional implications are very attractive.

National foundations have well-defined philanthropic goals.

There is usually fierce competition for these grants.

### Special Interest Foundations

Special interest foundations have a single field of interest.

They are not limited to a particular geographic area in their support.

They can be sources of special subject information as well as financial support. For example, the Joseph P. Kennedy Foundation provides information on mental retardation, the Robert Wood Johnson Foundation provides information on health, and the DeWitt Wallace/Reader's Digest Foundation provides information on education and libraries.

### Family Foundations

Most U.S. foundations are family foundations, such as the Brown Foundation, the McKnight Foundation, and the Evelyn J. Daniels Foundation.

The support pattern of family foundations is often motivated by personal interests; there is often no specific focus.

Family members often control the foundation.

Most of these foundations are very small, with no staff and a very limited geographical pattern of giving.

### Community Foundations

Community foundations, such as the Baton Rouge Area Foundation and the Cleveland Foundation, are not grant foundations but public charities.

Community foundations maintain their favored tax status by collecting money from the public and directing grants within the community for which they are named.

## Identifying a Prospective Funder

When you begin your research, you will discover thousands of potential funders for any given project. Obviously, it is impossible to appeal to every funder that might have resources for you. The following key questions and research parameters can help narrow down your choices and pinpoint a handful of funders who would be favorably disposed toward your particular project.

### Key Questions

1. Which groups, companies, foundations, or individuals have a vested interest in your project or activity? What kind of assistance are they most likely to provide: funding, in-kind support, advice on whom to approach?
2. Can board members or other volunteers be enlisted to assist in your fund-raising efforts by researching literature, identifying contacts, and so on?
3. Which local and national corporations and foundations are funding projects with objectives similar to yours?

### Research Parameters

1. Examine the funder's areas of interest. Do not look at education or health donors only—be creative!
2. Note the type of support given—is it basically in-kind?
3. Identify organizations and projects the funder has supported in the past.
4. Identify geographic and programmatic limitations associated with the funder. Pay particular attention to these limitations. Some foundations fund only programs located in certain areas of the country; corporations may fund only projects housed in the communities where they have facilities.
5. Determine the average gift from the funder.
6. Know and meet the funder's deadlines.
7. Identify contacts in prospective companies and foundations who are known to you or your board members.

## Writing a Grant Proposal

Once you have identified a prospective funder, it is time to write a grant proposal to request funding. Many people who have never written a grant proposal before are reluctant to try, but in large part, a successful grant proposal is simply a matter of organization. Be sure to consult the proposal materials from each individual funder to determine the areas to emphasize; requirements differ with each funder.

To begin, you may want to develop an action plan that will help give your project the proper focus and organization. The following guidelines for writing an action plan along with the accompanying sample action plan should help you formulate your own.

## Action Plan Guidelines

Your action plan should contain the following information:

**Who** is the target audience for your proposed effort? Provide a few details about your target audience. Who will be involved in planning the project? Try to involve as many key people as possible: principals, teachers, students, parents, volunteer scientists.

**What,** specifically, do you intend to do for this audience? Describe in this section what objective you wish to accomplish. Briefly describe the need that you plan to address and how this relates to science, math, and technology strategies.

**Where** will this proposed activity occur? Will the activity take place in the library or at some other site?

**How** will you accomplish your objective; that is, what is your plan of action? Describe the methods you will use to plan and carry out the project. Include activities, staff or volunteers who will be involved, materials to be purchased, and so on.

**When** will the proposed activity take place? Include the date(s) and time(s).

**Expected outcomes:** How will your project help to meet the need(s) you described? How will you determine what, if any, effect your program has had on the target audience?

**Budget:** In addition to the grant, will you use any supplemental funds from other sources (e.g., PTA, Friends of the library) for the proposed activity? If so, include these funds in your budget. How will the money be used?

**SAMPLE ACTION PLAN**

## Science Olympics

The following sample action plan describes how a $500 mini-grant will be used at the McNamara Elementary School Library. The grant will be used to stage a Science Olympics.

**Who:** The target audience for the grant is all third- through sixth-grade students in the school—approximately 300 students in fifteen classrooms. Students at McNamara School represent a cross section of the community. Approximately 65 percent of the student body is black, 25 percent is white, 7 percent is Hispanic, and the remainder is Asian. The student body has performed above the average in district standardized testing in reading and math. The school has an active parent/teacher organization whose members will be asked to volunteer in staging the Science Olympics. This idea has been discussed with the teachers and principal. They have expressed their support, and the teachers are willing to integrate the Science Olympics into their classroom science and math activities.

**What:** The objectives of this project are: (1) to raise the interest level of students and teachers in science; (2) to develop teamwork and problem-solving skills; and (3) to show the students that science can be fun. We hope to capitalize on the interest in the Olympics and to make this an annual event at the school. The Science Olympics will consist of four "events":

*Facts and Figures:* Teams of four students will attempt to solve a science puzzle using clues found in reference materials, books, periodicals, and other media in the library. The objective of this event is to encourage students to use a variety of science, math, and technology materials.

*Flights of Fancy:* Teams of students will build and fly paper airplanes. The planes will be judged on design and length of flight. The objective of this event is to introduce students to some basic principles of aerodynamics and design.

*Bridges of Understanding:* Using straws, tongue depressors, balsa, or other available materials, teams of students will design and build bridges that will be tested during the Olympics for their load-bearing capability. The objective of this event is to introduce the students to some basic principles of engineering and the use of geometric shapes in design.

*Great Eggspectations:* Using commonly available packaging material, teams of students will construct a container for an egg that will allow the egg to survive unbroken from a drop of ten feet. The objective of this event is to help students develop problem-solving skills.

There will be two levels of competition: Junior Olympians (third and fourth graders) and Senior Olympians (fifth and sixth graders).

The 300 students will form seventy-five teams, with each team having four students. Classroom teachers will assist in forming the teams for their classroom. Each team will enter only one event. This means that there will be approximately eighteen to twenty teams per event. Each classroom will select a country of the world to represent. As part of their preparation for the Science Olympics, each class will identify and learn about a famous scientist or mathematician either from that country or from an ethnic group represented in that country, or an American scientist whose ethnic origins relate to that country.

Medals or ribbons, depending on cost, will be awarded to each member of the first- through fifth-place teams. The remainder of the students will get certificates of participation. Judges will include teachers, parents, and a volunteer scientist for each event.

**Where:** The Science Olympics activities will take place in the library, the cafeteria, and selected classrooms.

**How:** The following activities will be undertaken to plan and conduct the Science Olympics:

- Brief the principal and teachers and secure their commitment to participate.
- Publicize event to parents (letter home) and students (flyers, student-created posters).
- Develop guidelines for each event.
- Secure judges.
- Purchase and distribute materials needed for the four events and order additional science, math, and technology materials for the library.

- Set up event areas.
- Conduct Olympics and present awards.
- Conduct a brief survey of selected students and teachers to see how they liked the event and what they believe they learned from it.

**When:** The Science Olympics will be held during National Science and Technology Week (April 24–30). The possibility of holding the actual event in the evening will be explored to allow parents the opportunity to assist or observe.

**Evaluation:** A brief survey will be conducted of selected students and teachers to see how they liked the Science Olympics and what they believe they learned from it. Classroom teachers will give students a writing exercise the day after the event asking them to write about what they liked most, what they learned, and what suggestions they have for the Science Olympics for next year. Students will be told that the librarian will use their essays to evaluate the project and to plan next year's Science Olympics.

**Budget**

| | | |
|---|---|---|
| *Funds* | | |
| Grant | $500 | |
| McNamara PTA | $100 | |
| Total: | | $600 |
| *Expenses* | | |
| Books and other science, math, and technology materials to supplement collection and for use in Facts and Figures event: | $350 | |
| Materials for event kits: | $150 | |
| 20 plane kits | | |
| 20 bridge kits | | |
| 20 egg drop kits | | |
| Awards and Certificates | $100 | |
| 40 medals or ribbons | | |
| certificates | | |
| Total: | | $600 |

## Fund-Raising Resources

Many publications list foundations and other sources that are interested in receiving grant proposals. The following are just a few suggestions to get you started, but we hope they demonstrate that there are many places "out there" that take an active interest in learning and libraries. And that's where you come in! Your ideas for enhancing the learning experiences of your students coupled with the finances and backing of a grant foundation could make for a winning combination for your students, your school, and your community.

### Corporate Publications

*The Corporate 1000—Yellow Book*

*Corporate 500: The Directory of Corporate Philanthropy*

*Taft Corporate Giving Directory*

*Standard and Poor's Register,* Vol. 1, *Corporations;* Vol. 2, *Directors and Executives*

*Directory of Corporate Affiliations*

*National Directory of Corporate Giving*

*Directory of International Corporate Giving in America*

Corporate Annual Reports

### Foundation Publications

*Foundation Directory*

*Source Book Profiles*

*Foundation Grants Index*

Foundation Annual Reports—Grant Guidelines

### Magazines and Newspapers

*Foundation Giving Watch* (newsletter)

*Donor Briefing* (newsletter)

*Corporate Philanthropy Report* (newsletter)

*Philanthropic Digest* (newsletter)

*Chronicle of Philanthropy*
*Chronicle of Higher Education*
*Forbes*
*Fortune*
*Business Week*
Local business publications

### Other Publications

*Federal Register*
Membership rosters of local business organizations
Foundation Center research services
DIALOG database service

## Possible Funding Sources

Listed below are some national offices and foundations that frequently award funding to schools and libraries. You can write or call them to receive guidelines for submitting proposals.

**ACS Teacher Travel Grants**
Office of High School Chemistry, American Chemical Society, 1155 16th Street NW, Washington, DC 20036, 202-872-4590

Applicants (high school science teachers) must design presentations that focus on American Chemical Society (ACS) programs. They may use grants to attend regional and national meetings of ACS and the National Science Teachers Association.

**Annenberg/CPB Math and Science Project**
901 East St. NW, Washington, DC 20004-2006, 202-879-9658

Projects should (1) create visual resources that will show how various teachers implement science instruction changes in kindergarten through eighth grade, (2) link rural math and science teachers by computer, (3) develop and implement new, cost-effective strategies for educating the public and parents, or (4) develop strategies for information and resource-sharing among teachers.

**Bristol-Meyers Squibb Foundation**
345 Park Avenue, New York, NY 10154, 212-546-4566

Projects in K–12 public education in the areas of math, science, and health education and the shortage of qualified teachers in those areas.

**Nathan Cummings Foundation**
885 Third Avenue, Suite 3160, New York, NY 10022, 212-230-3377

Projects in arts, environment, and health education.

**Du Pont Community Initiatives**
Du Pont de Nemours and Co., Du Pont External Affairs, N-9541, Wilmington, DE 19898, 302-774-6376

Elementary and secondary schools.

**Hasbro Children's Foundation**
32 W. 23rd Street, New York, NY 10010, 212-645-2400

Prefers projects serving poor and at-risk children.

**Edward G. Hazen Foundation**
505 Eighth Avenue, New York, NY 10018, 212-967-5920

Prefers programs involving disadvantaged and minority youth, partnerships, school reform.

**William Randolph Hearst Foundation and the Hearst Foundation, Inc.**
888 Seventh Avenue, New York, NY 10106-0057, 212-586-5404

Prefers disadvantaged and minority youth, private education, and scholarships.

**Hitachi Foundation**
1509 22nd Street NW, Washington, DC 20037-1073, 202-457-0588

Particular interest in collaborative projects involving schools and higher education institutions, museums, arts agencies, or other organizations; likes to see efforts that can be replicated in other communities. Hitachi is also interested in parent-involvement projects.

**Intel Foundation**
5200 N.E. Elam Yound Parkway, Hillsboro, OR 97124-6497, 503-696-2390

Projects enhancing math and science literacy.

**International Paper Company Foundation**
2 Manhattanville Road, Purchase, NY 10577, 914-397-1500

Prefers advanced students, at-risk students, curriculum development, disadvantaged and minority youth, educator training, partnerships, public education.

**The J. M. Foundation**
60 East 42nd Street, Room 1651, New York, NY 10165, 212-687-7735

Prefers community programs, economics, public education, volunteers.

**W. Alton Jones Foundation**
232 East High Street, Charlottesville, VA 22901, 804-295-2134

Prefers leadership development, partnerships, research, school reform.

**W. K. Kellogg Foundation**
1 Michigan Avenue East, Battle Creek, MI 49017-4058, 616-968-1611

Supports projects to develop new approaches in science education; to promote effectiveness and leadership in those who work with youths; and to provide child-care services.

**Kmart Corporation**
3100 W. Big Beaver Road, Troy, MI 48084, 313-643-1000

Supports innovative elementary and secondary education programs, research and curriculum development related to business or marketing, and programs dealing with child welfare and the family unit.

**John D. and Catherine T. MacArthur Foundation**
140 South Dearborn Street, Chicago, IL 60603, 312-726-8000

Basic skills, curriculum development, and educator training.

**The Andrew W. Mellon Foundation**
140 East Street, New York, NY 10021, 212-838-8400

Prefers college preparation, educator training, literacy programs, partnerships, pregnancy prevention, private education, and public education.

**Meyer Memorial Trust**
1515 SW Fifth Avenue, Suite 500, Portland, OR 97201, 503-228-5512

Projects in parent education, early childhood development, improving early educational opportunities, and new and effective ways to intervene with youth at risk.

**General Mills Foundation**
P.O. Box 1113, Minneapolis, MN 55440, 612-540-4662

Curriculum improvement, in-service teacher training, arts education, dropout prevention, and multiculturalism.

**Monsanto Fund**
800 North Lindberg Boulevard, St. Louis, MO 63167, 314-694-4596

Capital projects, public education, science and math, start-up programs.

**Charles Stewart Mott Foundation**
1200 Mott Foundation Building, Flint, MI 48502-1851, 313-238-5651

Curriculum development, disadvantaged and minority youth, family involvement, pregnancy prevention, preschool programs, public education.

**The David and Lucille Packard Foundation**
300 Second Street, Suite 200, Los Altos, CA 94022, 415-948-7658

Prefers at-risk students, basic skills, child care, disadvantaged and minority youth, family involvement, fine arts, job skills training, leadership development, partnerships, pregnancy prevention, scholarships.

**The Pew Charitable Trusts**
Suite 501, 3 Parkway, Philadelphia, PA 19102-1305, 215-568-3330

Prefers community involvement, disadvantaged and minority youth, dropout prevention, employment programs, family involvement, pregnancy prevention, substance-abuse prevention.

**Public Welfare Foundation**
2600 Virginia Avenue NW, Washington, DC 20037, 202-965-1800

Supports public school districts, higher education institutions, and nonprofit organizations that provide direct services to low-income and disadvantaged populations.

**The Rockefeller Foundation**
1133 Avenue of the Americas, New York, NY 10036, 212-869-8500

Prefers at-risk students, basic skills, arts and humanities, partnerships, cross-cultural activities.

**The Spencer Foundation**
875 North Michigan Avenue, Chicago, IL 60611, 312-337-7000

Prefers disadvantaged and minority youth, research, school reform.

**Texaco Foundation**
2000 Westchester Avenue, White Plains, NY 10650, 914-253-4000

Supports initiatives to promote math and science literacy as well as teacher and leadership training, in regions where Texaco has a significant presence.

## Funding References Especially for Librarians and Educators

### General References

*Annual Register of Grant Support.* Wilmette, Ill.: National Register, 1991.

Bauer, David G. *The "How To" Grants Manual: Successful Grantseeking Techniques for Obtaining Public and Private Grants.* 2d ed. New York: American Council on Education, 1988.

Burlingame, Dwight D. *Library Fundraising: Models for Success.* Chicago: American Library Association, 1995.

Greever, Jane C., and Liane Reif Lehrer. *Guide to Proposal Writing.* New York: The Foundation Center, 1993.

Lehrer, Liane Reif. *Writing a Successful Grant Application.* 2d ed. Boston: Jones and Bartlett, 1989.

Maolin, Judith B. *The Foundation Center's User Friendly Guide: Grantseeker's Guide to Resources.* 2d ed. New York: The Foundation Center, 1992.

Smith, Craig W., and Eric W. Skjei. *Getting Grants.* New York: Harper, 1980.

### Foundation Sources

*The Foundation 1000, 1998–99 edition.* New York: The Foundation Center, 1999.

This publication contains the most comprehensive information available on the 1,000 wealthiest foundations in the country.

*Foundation Grants Index: A Cumulative Listing of Foundation Grants.* New York: The Foundation Center, 1999.

Grant descriptions are divided into twenty-eight broad subject areas, such as health, higher education, and arts and culture. Within each broad field, grants are listed geographically by state and alphabetically by name.

*The Foundation Center's Guide to Grantseeking on the Web.* New York: The Foundation Center, 1999.

This is a complete guide to funding research online.

Haile, Suzanne, ed. *Foundation Grants to Individuals.* New York: The Foundation Center, 1991.

Jones, Francine, ed. *Source Book Profiles: An Information Service on the 1,000 Largest U.S. Foundations.* 2 vols. New York: The Foundation Center, 1991.

*National Data Book of Foundations.* 2 vols. New York: The Foundation Center, 1990.

## Corporate Sources

*Corporate 500: The Directory of Corporate Philanthropy.* San Francisco, Calif.: Public Management Institute, 1990.

*National Directory of Corporate Giving: A Guide to Corporate Giving Programs and Corporate Foundations.* 3d ed. New York: The Foundation Center, 1993.

Renz, Lorens, ed. *Corporate Foundation* Profiles. 6th ed. New York: The Foundation Center, 1990.

## Government Sources

*Catalog of Federal Domestic Assistance.* Washington, D.C.: Office of Management and Budget; U.S. General Services Administration, 1998. Annual.

*Federal Grants Management Handbook.* Washington, D.C.: Grants Management Advisory Service, 1988.

*Government Assistance Almanac, 1990–91: The Guide to All Federal Financial and Other Domestic Programs.* Detroit, Mich.: Omnigraphics, 1990.

*Guide to Federal Funding for Education.* 2 vols. Washington, D.C.: Education Funding Research Council, 1991.

## Specialized Sources

Blum, Laurie. *Free Money for Mathematics and Natural Sciences.* 2d ed. New York: Paragon House, 1987.

*Directory of Grants in the Physical Sciences.* Phoenix, Ariz.: Oryx, 1988.

*Grants for Libraries and Information Services.* New York: The Foundation Center, 1993.

Keats, Elizabeth. *Guide for the Preparation of Grants and Proposals for Starlab Portable Planetariums.* 1992. Learning Technologies, Inc., 59 Walden Street, Cambridge, MA 02140. Free.

*The National Guide to Funding for Elementary and Secondary Education.* 2d ed. New York: The Foundation Center, 1993.

*The National Guide to Funding for Libraries and Information Services.* 2d ed. New York: The Foundation Center, 1993.

## Periodicals

*The Chronicle of Philanthropy: The Newspaper of the Non-Profit World.* Biweekly. 1225 23rd Street NW, Washington, DC 10003-3026. Also online at http://philanthropy.com/.

*The Foundation Grants Index.* Bimonthly. New York: The Foundation Center.

## Online Services

DIALOG Information Services, Inc., supports three major files for locating foundation funding:

*The Foundation Directory* (File 26 on DIALOG) covers over 5,000 independent, corporate, and community foundations with assets of $1 million or annual grants totaling $100,000 or more. Provides detailed information on the foundations, giving interests and restrictions.

*The Foundation Grants Index* (File 27 on DIALOG) includes more than 300,000 actual grant descriptions for grants of $5,000 or more that have been reported by approximately 500 foundations.

*National Foundations* (File 78 on DIALOG) covers all 25,000 active U.S. foundations reporting to the IRS. Information is brief.

Dialcom is available to nonprofit organizations through the Telecommunication Cooperative Network, (212) 714-9780.

## Video References

*Introduction to Grant Funding.* Nonprofit Resource Center, Sacramento Central Library, 828 First Street, Sacramento, CA 95814.

## Internet Resources

*Council on Foundations*
http://www.cof.org/index.html

This site includes a glossary of foundation terms; press releases and current information about grants and funding; publications catalog; and more.

*The Foundation Center Online*
http://fdncenter.org/

A virtual gateway to philanthropy on the Web, this site includes news digests; the Foundation Finder, a search engine that contains approximately 46,000 records; an online proposal-writing short course; and an electronic reference desk called the Online Librarian. The Foundation Center also offers proposal-writing seminars that are held throughout the country. You can find a schedule and details on the Web.

*Grants Web*
http://web.fie.com/cws/sra/resource.htm

Here you'll find an excellent resource that contains links ranging from writing tips to news services to funding sources. Though geared toward the research community, the site contains something for everyone.

*GuideStar*
http://www.guidestar.org/

This is a Web clearinghouse of information on nonprofit organizations.

*TGIC–The Grantsmanship Center*
http://www.tgci.com/

Visitors to this site will find grant research information, and links to information about federal, state, and community foundations.

# 8

# Partnerships to Promote Science Activities in the Library

MARIA SOSA

Children are in school about six hours a day, five days a week, thirty-six weeks a year, but learning does not confine itself to this schedule. Children are learning all the time. Working as partners, families and other concerned adults in the community can play an important role in making sure that science and mathematics are part of what children learn every day. Libraries can help to initiate and nurture these partnerships.

To truly change science education, librarians, teachers, administrators, policy makers, parents, and community members must all contribute to the process of reform. The American Association for the Advancement of Science (AAAS) Science Library Institute sought to broaden opportunities for librarians to form partnerships with other stakeholders who share the vision of promoting excellent and effective education for all children. As the project descriptions in chapter 9 indicate, successful library programs result when librarians forge such partnerships with institutions and individuals outside the library. This chapter focuses on three key areas of partnership: school, home, and community.

## Partnerships in the School

Although classroom teachers have the primary responsibility for teaching science in the schools, librarians can help in a variety of ways. They can provide teachers with resources and can augment science instruction by providing informal science activities in the library. When librarians and teachers work toward a common goal, children will be the winners.

The teacher in a reformed mathematics and science curriculum is not merely a dispenser of information; the science teacher must become a facilitator of learning. This means that the teacher must create a rich, learner-centered environment in which students can actively pursue knowledge, primarily through inquiry, experimentation, and problem solving.

To create such inquiry-based environments, teachers must become familiar with resources, curriculum materials, and human resources (such as community members with special expertise) that support such learning. Librarians can provide the professional support teachers need by disseminating information about recommended resources and research findings. Librarians can be particularly helpful in providing materials that support students' explorations, such as children's tradebooks and magazines, videos, films, and computer software. Because librarians are often very comfortable and familiar with technology, they can also play a leadership role in providing technology training for teachers and school personnel.

Librarians can also support teachers' efforts by providing informal science activities in the library. Many librarians in the AAAS Science Library Institute conducted projects with a hands-on science component as part of their training. (See project descriptions in chapter 9.) Reports on these projects indicated that the librarians worked closely with science teachers to ensure that supplemental activities in the library supported ongoing instructional objectives. Such partnerships were facilitated by the training in science reform issues and pedagogy that the librarians received in the Institute.

AAAS Science Library Institute participants collaborated with teachers in many ways. At Paul VI High School in Fairfax, Virginia, school media specialist Diane Schule was able to demonstrate the value of the library to science teachers by making it a focal point for previewing science resources for classroom use. In the Duke Ellington School of the Arts in Washington, D.C., librarian Patricia Bonds engaged teacher participation by relating her Science Library Institute project to one of the objectives required for completion of senior high school biology. Media specialist Pauleze Bryant supported an integrated science curriculum by coordinating a joint effort that linked the chemistry and social studies classes at Calvin Coolidge High School in Washington, D.C.

For many librarians, the most difficult part of forging a partnership with teachers is taking the first step. In an already overcrowded day, librarians and media specialists may feel some reluctance to take on what might be perceived as additional responsibilities. The same is true of teachers. Science Library Institute participants found, however, that when both partners come together with a clear understanding of roles, the collaboration is productive and beneficial. Science Library Institute participants also actively enlisted support from school and library administrators. This support resulted in a deeper level of commitment.

Other important partners in the schools are principals and district administrators. Sometimes the school media center gets overlooked when schools are planning reforms or restructuring. One way to make sure this doesn't happen is to maintain contact with administrators at your school or school district. Sometimes this contact can be as simple as regular updates on activities in the library. In fact, sharing success stories with decision makers is a good way to not only build their confidence in what you have accomplished, but also interest them in providing additional support and resources for library projects.

## Partnerships in the Home

If families view science as an important subject, they will be more likely to promote science activities for their children. Libraries can help make parents more aware of the importance of science by providing opportunities for families to participate in informal science experiences.

Research tells us that parents play a vital role in the science and mathematics education of their children. Although most parents are willing to assume a more active role, many need to find an entry point. Libraries can help provide that entry point by sharing with parents a variety of strategies they can use to help children succeed in science and mathematics.

In 1996, AAAS conducted a review of educational research that looked at the role of parental involvement in children's education. Key findings from that review included the following:

1. Learning resources in the home, such as books and microscopes, have a positive impact on academic achievement. Too much television viewing has a negative impact.
2. High-achieving children spend a lot of time outside school in constructive learning activities, especially when encouraged by their parents.
3. Children whose parents set high standards show higher levels of academic achievement.
4. Children whose parents are directly involved in their school instruction, such as homework, perform at higher levels than children whose parents are not involved.

Librarians can encourage dialogue between parents and teachers. When parents and teachers communicate, children's classroom experiences can improve dramatically. Teachers can benefit from a parent's insight into a child's behavior and experiences. Parents can better understand the teacher's goals and help to support classroom learning. At Turner Elementary School in Washington, D.C., Institute participant Judy Bullock organized a science media fair that brought together teachers, students, and parents to view new science resources and participate in hands-on science activities. At Calvary Lutheran School in Silver Spring, Md., library media specialist Violet Lentner spearheaded a sharing-science activity during spring open house.

To encourage informal science experiences for children, libraries can provide parents with a list of community and science organizations that offer tutorial assistance and enrichment programs. Libraries can also disseminate information about activities at zoos, museums, nature centers, and parks, and about children's organizations, such as 4-H clubs. Librarians can help families do science and mathematics activities at home by creating science boxes for children to check out that contain simple activities or by putting together simple monthly science activities. To encourage families to share their results with the library, incorporate their projects into displays.

Many of these strategies can be inaugurated with a family night at the library. Science Library Institute participant Rose Pringle, of Roper Middle School in Washington, D.C., framed her project around a family event in order to gain parental support for Roper's new focus on mathematics, science, and technology.

## Partnerships in the Community

Every school or library is part of a larger community that has an important stake in making sure its children succeed in school. When educators reach out to the community, they often find many individuals who will gladly volunteer time and resources to enhance educational opportunities for children. Scientists and science-based organizations often have a special interest in partnering with schools.

Classroom or library visits and demonstrations by scientists can stimulate young people's interest and curiosity. An enthusiastic researcher can convey the excitement of making discoveries. An engineer can provide insights on solving real-world problems. And a mathematician can vividly demonstrate how mathematics describes and models scientific and engineering concepts. Scientists can bring librarians, teachers, and students up-to-date on the latest developments in their fields as well as describe career opportunities from a unique perspective.

Students and scientists alike can benefit from conversations in informal settings, such as libraries. During informal conversations, children can pursue their interests and questions about science with someone who really does it every day. These personal interactions not only can help students understand what makes scientists tick, but also give scientists better insights into the diversity of young people in our schools. Scientists can find out firsthand about students' concerns and thoughts the physical and social environments of today and tomorrow. Conversations with students give scientists an opportunity to expand their appreciation of how young people learn science and mathematics. Workshops conducted as part of the Institute also helped create partnerships between libraries and the community. Sharon M. Henry and the staff of the Agnes Parr Resource Center conducted two workshops to encourage home-school parents and Indian Education tutors to use hands-on science activities. Shari O'Keefe of Rapid City, South Dakota, conducted a health fair that brought together three elementary schools for an evening of learning in which fourth graders coordinated activities at seven different stations. Approximately 300 people attended the fair!

When scientists visit classrooms or libraries, they should be encouraged to conduct simple hands-on science activities with students. In this way, students engage in the phenomena of science and encounter the products of engineering through activities and projects that interest them. Students experience science as a way of knowing and engineering as a way of designing. By working with scientists, students can more readily acquire the spirit and meaning of science.

As part of her library project, Rapid City Institute participant Judy Gram recruited six guest speakers to attend the Dakota Middle School's "Discovering the Black Hills" program. The speakers were from the South Dakota State Archaeology Research Society, Forest Products Distributors, and the Inter-Tribal Bison Consortium and included a Lakota mentor and two rock climbers.

Another exciting Rapid City event sponsored by the AAAS Science Library Institute was Library Day at a Rapid City shopping mall. This turned out to be an ideal location—many shoppers stopped to look at the posters that displayed information about the local librarians' projects and to learn more about science reforms in their local schools. Local authors were on hand to autograph their books, and yo-yos, books, tangrams, and other items were given away free to children.

Librarians can also encourage teachers to take students to local scientific or engineering enterprises by disseminating information about field trips and tours. Students can see where science and engineering are done and obtain glimpses of what kinds of work scientists do. Workplace visits allow students to experience science careers firsthand as they talk with and watch scientists, engineers, and mathematicians in the real world.

At an AAAS Science Library Institute training workshop, program evaluator Mary Chobot led a brainstorming session that yielded a list of suggestions for using the skills of scientist partners in the library. Scientists can:

- assist in evaluating print and software collections for weeding and replacement;
- act as resources for simple hands-on experiments;
- attend brown-bag lunches with teachers;
- work with students as mentors;
- conduct seminars to generate ideas for science-fair projects;

provide information and assistance online;

advise on stand-alone science activities;

recommend media for circulation;

provide assistance in setting up science projects in a public library setting;

help librarians study the science of art and generate authentic murals (ocean or jungle scenes);

discuss career possibilities for minority and women scientists;

empower kids as scientists;

help librarians learn more about a specific technology (CD-ROM, telecommunications, Internet);

establish a resource database of skills;

help librarians learn more about specific programs that use telecommunications technology;

help students construct scale models; and

help students plant a garden (for beautification as well as learning).

Partnerships are as varied as the individuals who form them. Sometimes partners are brought together by long-term goals, and sometimes the objectives are more immediate. The pieces may already be in place in some instances, and in others the groundwork must be laid. In any event, AAAS Institute participants demonstrated that librarians and media specialists are critical links in making such partnerships effective. At Chillum Elementary School in Hyattsville, Maryland, school media specialist Rose Jones tapped into an existing coalition of parents, teachers, community activists, and business leaders to help make Chillum a county-designated science, mathematics, and technology school. Working toward a more immediate goal, Eleanor Organ led a group of students, teachers, and volunteers from the Environmental Protection Agency in a campaign to clean up the front lawn and playground of Bowen Elementary School in Washington, D.C. Partnerships have played a key role in the success of all the projects conducted by AAAS Science Library Institute participants. By taking an active role in forming partnerships, libraries and librarians can provide a wealth of science-based learning opportunities for children.

## National Resources for Educational Partnerships

A number of national programs and coalitions support community partnerships to improve K–12 education. A few are described here.

*The Center on School, Family, and Community Partnerships at Johns Hopkins University*
http://www2.ncup.org/us/intro.htm

The mission of this center is to conduct and disseminate research, development, and policy analyses that produce new and useful knowledge and practices to help families, educators, and members of communities work together to improve schools, strengthen families, and enhance student learning and development. The center publishes a variety of reports and bibliographies, abstracts of which are available at its website.

*The Education Alliance for Equity and Excellence in the Nation's Schools*
http://www.brown.edu/Research/The_Education_Alliance/

Based at Brown University, this organization responds to the needs of diverse student populations in the public schools. Believing that language, culture, and diversity are fundamental to the success of educational reform, the alliance creates partnerships with educators, policy makers, researchers, and business and community agencies.

*Latino Partnerships Pathway*
http://eric-web.tc.columbia.edu/pathways/latino_partner/

Visit this website to learn more about the cooperative efforts of universities, businesses, national organizations, and community groups that are working together to build a supportive network of partnerships for Latino students, their families, and schools.

*The National Center for Urban Partnerships*
http://www2.ncup.org/us/intro.htm

The aim of this organization is to help underserved, urban students successfully complete baccalaureate degrees. The center achieves this mission by creating enduring, citywide partnerships to effect systemic change in the educational system. These partnerships include K–16 educators and representatives from the community, corporate, and political sectors. At the center's website, you can find a great deal of information about partnership programs throughout the country.

# 9

# Science Project Ideas for Libraries and School Media Centers

MARY CHOBOT

BARBARA HOLTON

MARIA SOSA

*Science for All Americans*, published in 1989 by the American Association for the Advancement of Science (AAAS), was a clarion call trumpeting the need to increase Americans' level of scientific literacy. This effort must begin in the K–12 science, mathematics, and technology (SMT) curricula. New standards and educational restructuring to support the goal must be applied school by school and community by community. Within each school and community, librarians and media specialists are well positioned both to support SMT reform and to serve as catalysts to bring the necessary resources and parties together. The project examples in this chapter provide tangible evidence of the effectiveness of motivated library media specialists in providing leadership for literacy-enhancing activities in science.

Many of the activities can be implemented with minimal funds. Each AAAS Science Library Institute participant received a $500 grant, most of which was used to improve SMT collections in the media center. If funds in excess of the library budget are needed to purchase supplies, materials, or equipment, you might interest the PTA or other organization in supporting the project. Or ask your principal or library administrator whether the needed amount can be provided from discretionary funds.

To be successful, all the ideas, activities, projects, and events described here required some basic elements: planning; collaboration and coordination with teachers and others; enthusiasm; commitment to SMT reform; and a willingness on the part of the librarians to go the extra mile. The results speak for themselves. Young people who are turned on to science at an early age will become tomorrow's scientists and scientifically literate adults. We hope you find some ideas among those presented that will help spark an interest in science in your young library users.

The projects for this chapter were compiled and annotated by Mary C. Chobot, Ph.D, who served as a consultant and presenter for both AAAS Science Library Institutes, and by Barbara Holton, a participant in the Washington, D.C., AAAS Science Library Institute, who gathered information about science programs in libraries for the Office of Library Programs in the U.S. Department of Education.

## Elementary School Projects

The projects included in this section were developed and conducted by elementary school librarians or media specialists.

### Aviation Project/Super Plane Contest

Mary K. Pfeifle, library media specialist at Pinedale Elementary School in Rapid City, South Dakota, implemented this project with three fifth-grade classes over a six-week period. During week 1, students began keeping an Aviation Journal, in which they recorded their observations about what they learned as well as their overall impressions about the project. Students learned how to do subject and keyword searches using the online catalog and did searches for materials on aviation and paper airplane design. They were also introduced to the CD-ROM, *Jets*.

In the following weeks, the students made and flew paper helicopters, reviewed various paper airplane designs taken from sources identified in their online searches, practiced flying the different designs, discussed how design influenced the way the planes flew, and planned the design for the plane that they would use in the Super Plane Contest.

The contest took place during the project's fifth week. Winners were chosen based on the longest distance flown, and students selected the paper airplane with the best appearance and design. During week 6, prizes were given to the winners and the students completed their journals and participated in a wrap-up discussion.

To help students learn more about careers in engineering and aerial photography, the project included a field trip to an engineering firm that has used aerial maps to develop computer images of Mount Rushmore.

Mary reported that "it was wonderful to see the way the students embraced this project so enthusiastically, to see how many of the girls were as competitive and interested as the boys in what is still a male-dominated field, and to have another opportunity to work in a learning environment that makes me as a librarian a more active part of my students' elementary school life." Mary also served as the librarian at Lincoln Elementary School last year; this year, she serves three schools.

## Bowen Improvement Project (BIP)

At many inner-city schools, trash and debris on school grounds pose a threat to the safety and well-being of the children. The appearance of the Anthony Bowen Elementary School in Washington, D.C., had been an ongoing problem. Removal of broken glass and trash from the playground area required approximately two hours of janitorial time each day. This debris injured an average of three students during each recess period. Library media specialist Eleanor H. Organ decided to do something about the problem.

The objectives of her project were to clean up the immediate school environment (the playground and the front lawn), to recycle the materials collected, and to raise the interest levels of the teachers and students in the quality of their environment.

The project was conducted in three stages. During stage 1, students cleaned up the site by collecting the trash; sorted what they found into four groups—paper, glass, metal, and other recyclable materials; and recorded the amount of material in each category. Stage 2 emphasized building students' pride and involvement in the project. A presentation informed students about the amount and kind of trash found in the school's environment and the dangers posed by living and playing in such a harmful environment. Speakers urged students to become involved in keeping their school environment safe and clean. Stage 3 consisted of a plan for continued upkeep of the grounds. The process described in stage 1 was repeated in two-week intervals—trash was collected and sorted, and the results were compared with data from the earlier collections.

The project, which took place during May, involved the students, teachers, and others in the school community in several activities. The principal, the science teacher, classroom teachers, and Environmental Protection Agency (EPA) volunteers were briefed on the Bowen Improvement Project (BIP). Students and teachers viewed a video on the value of a healthy environment. Selected students attended a presentation by the EPA. Classroom visits were made. Forty fourth-grade students collected and sorted the trash, recorded and compared the results, and then graphed the results from each collection to show increases or decreases in the amount of each kind of trash found. Follow-up visits were made to classrooms to show these results to students and ask for their continued support. A party was held on the last cleanup day. Students were queried about their perception of the project, and, because of their enthusiastic support, the BIP will become an annual event. As a result of the project, students are taking greater pride in their school environment and their ability to help make it a safer place.

## Butterfly Explosion

The Butterfly Explosion was the result of a science project developed by Jessie K. Campbell, library media specialist at McGogney Elementary School in Washington, D.C. The project involved approximately 500 elementary school students from prekindergarten through sixth grade.

The objectives of this project were (1) to develop the concept that although butterflies are a universal insect, different species are found in different countries or regions of the world; and (2) to increase interest in science among students, with a special emphasis on the care of living things.

Each of the twenty-one classes at the school was assigned a country and was responsible for nurturing butterflies from that country. Larvae were purchased from a biological supply company. Each class prepared a glass-jar home for its butterflies, adding leaves and sticks to provide a natural environment. It took about three weeks for the butterflies to hatch. Once the butterflies had hatched, the students placed the glass-jar homes in a dark place or covered them with black construction paper, so that the butterflies would remain inactive until they could be released, thus protecting them from injury.

During the project, students observed the butterfly life cycle, specifically the maturation of the painted lady butterfly; identified butterflies as insects; graphed the stages of maturation of the insect; drew butterflies representing various countries; read and wrote stories about butterflies; conducted library research on the variety of butterflies worldwide; prepared the soil and planted plants to attract a variety of butterflies; cultivated and cared for the plants; developed observation skills; developed a concept of the interdependence of living things; and developed skills in caring for plants, butterflies, and other living things.

The project culminated with an outdoor assembly, during which all the butterflies were released in a Butterfly Explosion. The assembly program also featured class displays and was accompanied by music and festivities.

## Current Events, Books, and the Library

Jane Ostrand, school librarian at Kingsbury School in Oxford, Michigan, devised a project to increase students' awareness of the scope of materials in the library. Throughout the school year, she called attention to items from the collection by displaying newspaper articles along with books on the subject and by providing directional signs indicating where students could look for more information. For example, when beavers began building a lodge in a pond on school property, Jane displayed books that had information about beavers, including *Meet the Beaver* by Leonard Lee Rue III and *Busy Beavers* by M. Barbara Brownell.

In addition she highlighted any school, local, national, or international events that might inspire children to begin their own research. When a child brought a pet to school for show-and-tell, Jane gathered and displayed library materials about the care and feeding of pets. News reports that could lead to scientific inquiry included stories on oil spills, rain forests, alternative sources of energy, and wayward whales.

## Fascinatin' Physics

The four physics kits that Barbara Holton developed for Watkins Elementary School in Washington, D.C., provide an excellent means of introducing elementary school students to some basic physics concepts in an interesting, hands-on approach, while motivating them to read more about science topics.

The kits cover the following topics:

*Sound:* Several tuning forks, including an adjustable one, give students the opportunity to experiment with sound waves.

*Mirrors:* Concave, convex, and flat mirrors are used to demonstrate the properties of mirrors.

*Magnetism:* Using magnets of different sizes and shapes, a dowel rod, and iron filings encased in Plexiglass, students learn about the attracting and repelling power of magnets.

*Spectrums:* Prisms and diffraction gratings are used to examine the colors of natural and artificial light.

A booklet with the following elements accompanies each kit:

questions to answer

a hypothesis

an experiment, using materials contained in the kit

a procedure, with suggestions and questions

results

conclusions

further investigations, with ideas to try

a section titled Read More about It, containing a list, with call numbers, of books the library has on the topic

other subject headings to try if students wish to find more information on the topic.

The booklets, produced using WordPerfect, are nicely formatted on four pages and are laminated for protection.

Each kit also contains a logbook in which students can record observations, charts, graphs, notes, and so on. The kits are packaged in zip-lock plastic bags for easy storage and checkout.

The materials needed to create several kits on each topic cost approximately $150. The kits were used at the Watkins Elementary School with great success. Students worked in pairs, taking turns reading the directions and doing the experiments, and then recording the results in the logbook. Students enjoyed the experiments and expressed enthusiasm for these hands-on experiences.

## Health Fair, Human Fare, the Whole Affair

Shari O'Keefe, library media specialist, divides her time among three Rapid City, South Dakota, elementary school libraries: Garfield, Horace Mann, and Rapid Valley. Not wanting any of her 1,200 students to miss the opportunity to participate, Shari implemented her project for all three schools. The health fair was held at the centrally located Horace Mann Elementary School.

Fifty percent of the students at Garfield and Horace Mann are Native Americans; at Rapid Valley, a large percentage of students are transient because their parents are in the U.S. Air Force. Because an important objective of the project was to encourage parent participation, the fair was held in the evening.

The fourth-grade teachers agreed to integrate the health fair project into their curriculum, either as class instruction or individual projects. Although the fourth graders were the facilitators of the health fair, all grade levels from each school were encouraged to participate in the event.

Approximately 300 people passed through the fair and visited the seven stations, where fair-goers examined the brain sizes of different mammals; were given recipes for healthy eating and samples of healthy food; saw a demonstration of what smoke does to the lungs; had their blood pressure taken; exercised and had their pulse rate taken; role-played situational conflict resolution and thought of ways to relax; and observed students working with garbage and assessing how the planet will appear to future generations.

## Environments, Habitats, and Ecology

Patricia Munoz of Randolph Elementary School in Arlington, Virginia, initiated a schoolwide project to provide a variety of experiences related to environments and habitats for teachers and students in grades K–5. The objectives of the project were to (1) support the school program by providing hands-on science experiences, (2) teach children to use elements of the scientific method, and (3) incorporate information retrieval skills into the science program. The first-grade classes participated in story hours relating to various habitats and biomes. Second-grade classes studied rain forests and swamps. Two third-grade classes studied mammals and used reference books in the library to graph the life spans of common mammals. Some fourth graders did individual projects related to the overall theme of ecology. Fifth graders studied rain forests and the water cycle. The library assisted students with their research and provided materials to support the efforts of the students and teachers. Many of the student projects were displayed as murals throughout the building. Displays of student projects were also available for parent viewing during an evening showcase.

## Information Skills and Science

The purpose of Ann Potter's project was to help third-graders at McKinley Elementary School in Arlington, Virginia, develop information skills by using library materials. Third graders received instruction in information skills and science through cooperative planning and teaching between the classroom teachers and Ann, the library media specialist. During a yearlong series of lessons and activities, the students learned how to use an index, the phone book, title pages, the Dewey Decimal Classification System, and reference books. They also learned what primary and secondary sources are.

The science component of the program focused on classification skills. The students independently, watched an audiovisual program that described the difference between living and nonliving things. After the program, the students sorted objects pictured on large, numbered cards into living and nonliving groups and recorded their answers on prepared response sheets.

As a class, the third graders grouped themselves in the library media center according to various attributes: gender, eye color, hair color, number of siblings, and favorite colors. Each child was given a large picture card of an animal to hold, and they classified themselves as mammals, birds, reptiles, amphibians, and fish.

## Jack and the Beanstalk

The target audience for this project was a group of English as a Second Language (ESL) students from several ethnic groups, including Hispanic, Asian, African, Polish, Hungarian, and Italian. Because the students had limited English speaking and writing skills, Billie Branscomb, the librarian at Hearst Elementary School in Washington, D.C., devised a project involving photography to allow students to express themselves in a new way while learning some basic science concepts.

The science objectives of the project were to (1) help students understand that plants are crucial to life on earth; (2) observe the germination of a seed and its growth stages; and (3) measure plant growth to the nearest centimeter. Each student was given a set of hyacinth bulbs, some soil, and paper cups in which to grow their bulbs. Students used a camera to photograph their bulbs before and after germination. When the bulbs sprouted seedlings, the students photographed them every five days and recorded their growth on charts. On Arbor Day, April 29, students photographed the planting of their seedlings on the school's campus.

## Make a Difference . . . Reduce, Reuse, Recycle!

The objective of this project was to make students aware that throwing away recyclable paper affects the environment. Cheryl L. Brenden, library media specialist at Grandview Elementary School in Rapid City, South Dakota, felt that the students must actually get involved in recycling to fully embrace the idea. The entire school population got behind the paper-recycling project: 441 students in grades K–5, their teachers, and all staff members. Cub Scout Pack 53 and their leader, Jill Avey, also participated in the recycling program.

After communicating her plan to the principal, staff, students, and parents through bulletin boards, newsletters, Monday morning announcements, and lesson plans, Cheryl continued to encourage and educate. Books on recycling were displayed in the library, and the librarian led discussions on recycling, environmental issues, and our nation's trash dilemma. The coordinator of the "Keep Rapid City Beautiful" campaign spoke to the staff and students about the city landfill and had each student construct a "trash pizza" to illustrate what goes into a landfill and what to look for in a recyclable item.

The recycling program was very successful. At the end of the first week, the recycling bins weighed in at 161.5 pounds of paper, which is the equivalent of saving 1.4 trees. (One ton of paper saves seventeen trees.) By the tenth week, the students and staff had collected a total of 875 pounds of paper—7.4 trees had been saved! Cub Scout Pack 53 planted two trees in the school yard to celebrate the success of the recycling project. In just ten weeks, the students at Grandview learned that they *can* make a difference.

## Nature Classroom

This project created a permanent outdoor hands-on nature classroom for 420 K–5 students at Corral Drive Elementary School in Rapid City, South Dakota. Because the elementary school and Southwest Middle School are in the same building, the classroom will also serve 534 Southwest students in grades 6–8. The two librarians, Barbara Allum and Kathy Benson, worked together to develop complementary projects. Young flower plants resulting from the Southwest project were planted in the nature classroom. (See "Flower Power" later in this chapter.)

The nature classroom, located on a ponderosa pine–covered hill approximately 25 yards south of Corral Drive Elementary, is an ongoing, cooperative project. An outdoor thermometer, wind and rain gauges, a telescope, and a soil analysis kit were acquired for use in the nature classroom. A volunteer landscape architect helped with the planning for the project. The fifth-grade students built bluebird houses as a community project and donated one for the nature classroom. Vegetation produced in school science experiments will be planted in the nature classroom. Parent volunteers and other service organizations were enlisted to help complete a path and other aspects of the nature classroom.

The nature classroom's peaceful, outdoor setting provides an excellent venue for many activities, such as listening to guest speakers; stargazing and moonwatching; identifying native plants and animals and learning about their habitat; discussing the ecosystem; studying and comparing the life cycles of ants, birds, and other animals; studying climatic data, such as average temperature and precipitation; weather mapping; observing flora and fauna to develop a field guide for local plants and animals; determining tree trunk circumference, height, and spread of the crown; analyzing the soil; studying erosion and compaction problems; and increasing student awareness of environmental problems and how the students can be part of the solution.

## Passport to K–6 Student Success in Science, Science Awareness Fair

At the time of this project, Chillum Elementary School in Hyattsville, Maryland, was working toward becoming a science, math, and technology school. A team of parents, teachers, community activists, and business leaders had been sharing ideas and pursuing this goal. A committee consisting of library media specialist Rose Jones, the principal, two teachers, and a community member documented how this effort might be supported, especially by the library media center.

The committee decided to use Institute funds along with money generated by a book fair to purchase new science materials for the library media center. To showcase these new materials for parents, students, and other interested members of the community, a science awareness fair was held on a Saturday in May.

Several mini-workshops and science stations were set up in the library and throughout the school building. A NASA aerospace education specialist demonstrated how astronauts live and work in space. Students were able to see firsthand what a space suit looks like and learn what space food is like. Parents and students were able to ask questions, and one student got to try on the space suit. A station in the library displayed new science materials, played a video on careers in science, and housed student exhibits. A teacher volunteered to be a "teacher in space," much to the students' delight. At other stations participants could explore magnetism and structures.

The science awareness fair was well attended by both parents and students. So much enthusiasm was generated among teachers, students, and parents that another event was planned—an "Invention Convention." These activities have produced more teacher and parent cooperation, more volunteering by community leaders, the completion of a science laboratory in the building for grades K–6, and more enthusiasm and interest in science on the part of students.

## Physics Fair

At Capitol Hill Cluster School, Watkins Primary Campus, in Washington, D.C., library media specialist Cathy Pfeiffer coordinated a week-long physics fair during National Science and Technology Week in which one hundred fourth-graders participated. A new science laboratory was used for the first time during the fair. In the lab, the science teacher set up experiments and activities on light and sound for primary students and experiments on magnetism and electricity for intermediate-level students.

In the library media center, the physics kits on sound, mirrors, magnetism, and spectrums created by Barbara Holton (see "Fascinatin'

Physics") were available for students to use during their regular library visit. The library had also purchased kits, books, and other materials to support further investigation of these topics.

Students and teachers responded with much enthusiasm to the physics fair, which generated greater interest and involvement in science. After using one of the physics kits in the library, a fourth-grade student explained, "I felt good that I could explain in my own words what had happened."

## Science and Technology in Story Times

Badger Clark and Carousel Elementary Schools are located in the Douglas School System in Box Elder, South Dakota. The target group for this project was the kindergarten and first-grade students in these two schools.

The objectives of this story, experiment, and problem-solving program were to (1) raise the awareness and interest level of these students in science and math, (2) develop students' problem-solving skills, and (3) demonstrate the link between literature and science, math, and technology in everyday situations.

Library aide Nancy A. Eldridge selected children's books that would lend themselves to introducing science, math, or technology concepts. After the children heard each story, they engaged in a hands-on activity. Some of the stories and activities that were used effectively with these students are described here.

Chris Van Allsburg's Caldecott Medal–winning book, *Jumanji* (Boston: Houghton Mifflin, 1981), was the first story read to the children. To help the students understand the concept of chemical reactions, Nancy created a volcano using baking soda, vinegar, and soap. The children also created their own volcanoes.

*The Mousehole Cat* by Antonia Barber (New York: Macmillan, 1990) encouraged the children to learn about how electricity can generate light. Batteries, tiny light bulbs, and aluminum foil were used to create an electrical circuit. The children each completed a circuit and watched with surprise as the tiny bulbs lit up.

Paul Galdone's version of the story of *Henny Penny* (New York: Clarion, 1979; paperback, 1984) created an opportunity to ask this question: Can things really fall from the sky? Guest speakers were invited to talk to the children about such topics as tornadoes, hail, space junk, meteors, meteorites, stars, lightning, aircraft, and birds (falling out of the sky because of pollution). Each class chose a topic and researched it using books, the Internet, information from guest speakers, conversations with parents, and other personal knowledge.

*A Candle for Christmas* by Jean Speare is a lovely story about a young boy who fears his parents will not make it back to the reservation for Christmas. His love and concern for his parents, and the candle he lights, reunite them on Christmas Eve. This story made the children eager to conduct experiments with candles and to learn more about fire and air. A simple experiment demonstrated that a candle flame needs oxygen to burn.

Other books included *Thunder Cake* by Patricia Polacco, a story about a grandma who finds a way to dispel her grandchild's fear of thunderstorms. This book created an opportunity to study thunder and lightning, two weather phenomena that often frighten children. And *Avocado Baby* by John Burningham was used to introduce concepts related to health and nutrition.

To promote the project, Nancy created attractive flyers for some of the stories and accompanying activities. Many of the activities described here could also be used in public libraries during story hours for young children.

## Science Everywhere Fair

St. Elizabeth Seton School in Rapid City, South Dakota, serves 450 students in grades K–8. Librarian Jane Holeton staged an interactive science fair that involved the whole school and for which the middle-school students created hands-on science activities contained in decorated, color-coded bags.

The Science Everywhere Fair took place in classrooms, the library, the gym, and students' homes—everywhere. All content area teachers, across the curriculum, participated. Middle-school students researched topics they were interested in and put together bags containing an experiment, facts, an information file, a short biography, and so on. Creativity was encouraged, and the science teacher, the librarian, and other teachers were available to help the students as much as needed. When the contents for their bags were ready, the older students tested the activities and materials with the elementary-level children to see if any modifications were needed and made revisions as required.

On the morning of the fair, the completed projects were judged by community members using preestablished judging criteria and a point system. Fifty projects were awarded ribbons and displayed in the gym. In the afternoon, all K–5 students came to the gym to do the hands-on activities with the middle-school winners. Parents and community members were invited to view the activities.

All students received a grade in science, English, and art for their completed project. A grading sheet was developed and points assigned for various elements to assure consistency in grading. The fifty winning projects, in their colorful bags, were added to the Seton Library collection and can now be checked out for individual or classroom use.

## Science Olympics

Glenda B. Johnson, library media specialist, organized the Science Olympics at Robbinsdale Elementary School in Rapid City, South Dakota. Nearly two hundred fourth and fifth graders participated in this two-day event. Opening ceremonies began at 8:30 A.M. as eight teams marched in carrying the flag of the country they had decided to represent. Twelve events were held for the fourth grade, and sixteen events for the fifth. Activities and supervision were also provided in two video rooms, a science activities room, and the computer room, which students could use when they were not participating in events.

Students competed in the following events: Aerodynamics, Barge Building, Catapult, Geo Boards, How Do You Spell Science?, Naked Egg Drop, Naturally South Dakota, No Bones About It, Password, Pentathlon, Reflection Relay, and Rock Hound. Fifth graders competed in four additional events: Energy Box, Solar Collector, Sky Scrapers, and Science Bowl.

At the end of the day, medals were awarded to the first-place team (highest combined scores for all the events) in each grade level. A traveling trophy was awarded to one room in each grade level whose two teams had the highest combined scores. Every student received a placement or participant certificate for each event in which she or he participated.

## Vegetable and Flower Gardening

The objectives of the Vegetable and Flower Gardening Project at Lyles-Crouch Elementary School in Alexandria, Virginia, were to (1) raise the interest levels of students in gardening, (2) beautify the school grounds, (3) raise the self-esteem of students and their pride in school property, and (4) share the responsibility and maintenance of the vegetable and flower garden they cultivated.

Librarian Ruby Osia planned and implemented the project with Andrea Courduvelis, the fifth-grade teacher whose class of twenty-one students planted and cared for the garden; the physical education teacher; the custodian; and the maintenance workers at the school.

Students began to study gardening in March. Ruby and Andrea introduced the unit, which included library resources the students could use to conduct research on various gardening topics. The students found information on planting and maintaining a garden. They measured plots for two gardens, and the maintenance workers, the custodian, and students worked together to break ground and till the soil. Pansies were planted in early April, and tomato and cucumber plants were planted at the beginning of May. The students kept a journal of all the work they did and of their observations about plant growth. The project was well documented with photographs taken by the librarian.

Resource persons came to the school to share gardening techniques and information with students throughout the project. Students had the opportunity to interview a botanist, professional gardeners, and garden supply store owners.

Students were very enthusiastic about the project. They became more knowledgeable about gardening, and some of them expressed a desire to continue gardening at home. Students enjoyed harvesting and eating the vegetables they grew and took pride in the fact that they had helped to make the yard more attractive, pointing out their garden to many visitors to the school. Because several math and science concepts were integral to the project (for example, measuring the garden plots and learning how seeds develop into mature plants), the students learned that science can be both fun and useful.

## Middle School Projects

The projects included in this section were developed and conducted by middle-school librarians or media specialists.

## Discovering the Black Hills

Dakota Middle School in Rapid City, South Dakota, is a large, urban school serving 950 students in grades 6–8. The seventh-grade environmental science class, taught by Bill Casper, focuses on the Black Hills of South Dakota. The class studies the geologic, archeological, environmental, and multicultural aspects of the area. Important geologic and historical sites are identified, mining and timber concerns are explored, and Lakota traditions and beliefs concerning the land and its resources are presented.

The AAAS mini-grant funds were used to purchase books, films, CD-ROMs, and topographical maps of the Black Hills and South Dakota that will support the curriculum for this science class as well as the social studies curriculum. Library media specialist Judy Gram worked closely with Bill to select materials that would support the objectives for this six-week course.

Six guest speakers, including Michael Foscha from the South Dakota State Archaeology Research Society; Chris Legner from Forest Products Distributors; Ben Little Bear, a Lakota mentor; Andrew and Peter Gram, rock climbers; and Carla Brings Plenty, from the Inter-Tribal Bison Consortium, spoke to the students. These speakers addressed the importance of studying the people who lived here before us, the physics of hunting the mammoth, timber as a renewable resource, Lakota traditions and beliefs, the science behind rock climbing, the geology of the Black Hills, and the ecological importance of bison in North America. Students were very enthusiastic about this class, the new materials purchased with grant funds, and the guest speakers. Judy hopes that eventually the class can be offered in all five middle schools in Rapid City.

Judy has compiled a bibliography of print and nonprint materials, titled "Discovering the Black Hills." Readers who want to contact her about the project or the bibliography can reach her by mail or by e-mail at dakotalib@rapidnet.com.

## Endangered Species: Research and Awareness

Approximately 760 students in thirty-two classrooms, grades 5–8, attend Williams Middle School in Sturgis, South Dakota. The objectives of library media specialist Diane Bilbrey's project were to raise the awareness level of the students and community to the plight of endangered animals, to develop research skills, and to create a mural on the library walls as a lasting testimony to the endangered animals and to the students.

The mural is a reminder to everyone that awareness helps bring about solutions to a problem. A local artist volunteered her time to paint the mural after the students did the research and the preliminary sketches. Each class was assigned a specific endangered animal to research. The students were able to watch the painting in progress, and to see their animal become part of the mural in its specific habitat.

In an all-school assembly, a representative from the Forest Service gave a presentation about area animals that have become endangered. A cross-curriculum unit was developed involving teachers from several disciplines: In social studies, students learned about biomes; in language arts, they prepared oral and written reports on the animal of their choice; they developed library research and technology skills by using print and other media resources to gather information about the animals; in math, they used graphs, charts, and statistics; and in science, students learned what the animals eat, their habitat, and why they are endangered.

This was a cooperative effort, and the student council and book-fair profits provided significant additional funds for support. Williams Middle School now has a beautiful mural on the library walls—a reminder to everyone to protect these animals and a testimonial to the students' efforts.

## Family Science in the Library Night

Roper Middle School in Washington, D.C., became a math, science, and technology school for the 1993–1994 school year. Family Science in the Library Night provided an opportunity for parents to see how the library media center supported the school's new focus. Librarian Rose Pringle's project gave parents a chance to examine the library materials available to their children, to see how students used these materials in their learning activities, and to explore the technology used by their children.

In order to ensure the success of the event, the librarian briefed the principal and teachers and secured their commitment to participate.

Letters sent to parents and student-created posters and flyers publicized the event.

Guidelines were developed for each of the four program components:

We're So Proud displayed the year's winning science fair projects and the library resources used to help pull them together.

Slide Tape Presentation offered views from the classroom, links to the library, and statements from students and faculty about the math, science, and technology program. Resource and Technology Time presented descriptions of books and demonstrations of hardware and software available in the library. Hands-on! provided families the opportunity to try new software and to work with the computers their children use.

## Flower Power

Southwest Middle School in Rapid City, South Dakota, occupies the other half of the building where Corral Drive Elementary School is located. Working cooperatively, library media specialists Kathy Benson and Barbara Allum planned complementary projects. Southwest established a plant nursery in the media center to conduct research on plant and flower growth. The products of this nursery were then used to beautify the school courtyard and to provide plants for the outdoor nature classroom at adjoining Corral Drive Elementary.

Ten sixth-grade girls who were meeting with the school counselor to improve self-esteem and to develop social interaction and communication skills were a particular target for this project. While planting, tending to, and experimenting with the plants, the girls kept a journal of their observations; improved their team interaction, communication, and problem-solving skills; and developed a greater awareness of science- and math-related occupations. The other 525 middle-school students in grades 6–8, who regularly use the media center and enjoy the courtyard during their lunch hour, also benefited from the project.

Many individuals and groups cooperated in the Flower Power project, including the school's principal, counselor, art instructor, building custodian, and two librarians; a local plant nursery; and the local garden club. The National Agricultural Library's Plant Pal website was an excellent resource.

## One Hundred Scientists of Color

Edna Becton Pittmon, library media specialist at Langley Junior High School in Washington, D.C., devised this three-week experimental cooperative learning program in which students researched the contributions of outstanding scientists of color. Because many of the more than seventy students who participated were considered at risk, an important goal of the project was to build self-worth and self-esteem. The three ninth-grade science teachers, parents, an administrator, and the librarian worked together to plan the project.

The objectives of the project were to (1) provide activities to expose students to numerous examples of outstanding scientists of color, and (2) provide effective methods of cooperative learning that would contribute to students' social skills and their overall ability to interact successfully with others, thereby increasing their feelings of self-worth and self-esteem.

The project was divided into classroom and library segments. The librarian introduced students to resources they could use in their research, and teachers showed them how to develop charts, posters, graphs, and other ways to display or communicate the contributions of the scientists. Students were then organized into teams, and project leaders discussed with them the advantages of cooperative learning. The student teams used a variety of science and technology reference materials, books, videos, periodicals, and other media to conduct their research. Students demonstrated a great deal of creativity in presenting their findings: They made bookmarks, posters, charts, graphs, biographical listings, and even a book in the shape of a clock.

To support this project, the library media center used its AAAS Science Library Institute grant to purchase additional science resources, including reference materials, books, three science periodicals, a video, and other supplies.

The project had several beneficial outcomes. The science teachers and the librarian established strategies for exchanging ideas, communicating, and working effectively with students in hands-on science activities. The library media center acquired several needed science resources. Students expressed creativity, enthusiasm, a great sense of accomplishment, and a desire for more science-learning experiences. The project increased their science awareness as well as their self-esteem.

## Rocks and Minerals of the Black Hills

Jean Diedtrich, district librarian for the Custer School District in Custer, South Dakota, serves the needs of all K–12 students in this small district. Her project focused on teaching 80 eighth-grade science students at Custer Middle School about the geological formations in the area in which they live—the Black Hills—and demonstrating to students that rock and mineral identification can be fun.

A guest speaker from the School of Mines and Technology spoke to the students about rocks, minerals, and the geology of the Black Hills. The science teacher, Mrs. Witt, gave a presentation and created labs and assignments for the students to do. Both she and Jean were available to help the students with this project. Each student chose one rock or mineral to study in depth, then researched its mining formations and other facts in the media center, devised a lab experiment that demonstrated its properties, and created a visual display or chart showing where this rock or mineral is most commonly found. Several cases of rocks and minerals, which the students were required to identify, were on display in the library.

The students expressed their enthusiasm for these experiences through which they gained a greater appreciation for some of the wonders of the area in which they live. In the future, Jean hopes that this experience and the lessons learned from the project can be expanded to include the high school ecology class as well as the elementary school science classes.

## Super Sleuths, Inc.

Library media specialist Annette Ortiga of Jefferson Junior High School in Washington, D.C., wanted to stimulate students' interest in scientific investigation, strengthen their critical thinking and investigative skills, and show them that science can be fun. To do this, she provided an opportunity for the entire student body of 769 seventh, eighth, and ninth graders to participate in a Super Sleuths, Inc., training session.

To implement this project successfully, Annette met with the director of science and the principal of the school to get their input and support. All the teachers in the science department expressed a willingness to support the project and to have their students participate in it.

Students were presented with the following scenario: "Super Sleuths, Inc., employs a team of investigators whose job it is to seek answers to scientific questions. The investigators use information gathered from various reference tools and the results from hands-on science experiments to help them formulate valid conclusions. You've just been hired by Super Sleuths, Inc., to solve one of their most baffling cases. After you check in to your office, you will be handed an investigator's kit containing a data sheet for collecting information from selected science books, directions for doing an experiment, and materials for the experiment."

An example of the experiments is "Building Strong Bridges." Students are provided with several sheets of typing paper cut in half lengthwise, two stacks of books of equal height, and several pennies. They are asked to experiment with these materials to build the strongest bridge they can, trying to support as many pennies as possible, and to record the results.

The training sessions and investigations took place in the library media center. The project was planned for School Library Media Week in April. Students worked on their experiments in teams. To support the project, the library used Institute grant funds to purchase (1) several new science reference books and other print materials to update its science collection, (2) science magazine subscriptions, and (3) supplies for the experiments.

The project was well received by teachers and students. Students enjoyed doing the experiments and seeking out answers as Super Sleuths. After this experience, many students expressed more interest in science.

## Student Reviews of New Science Materials

Hardy Middle School in Washington, D.C., has a student body of 225 students in grades five through eight. The objectives of their project, devised by library media specialist Mary Jane Cox, were to upgrade the science book and software collection and to use the new materials to improve students' critical thinking and writing skills. The latter objective was accomplished by having students read the new acquisitions and then write their own reviews.

Students assisted with inventorying and weeding the science collection and, along with teachers, made suggestions about which materials needed updating. Final selections were based on teacher and student recommendations, published book reviews, and the librarian's research.

The new materials arrived fully processed, so they could be given to students immediately to review. Mary Jane and a science teacher developed guidelines for students to follow in writing their reviews. In addition to the usual bibliographic information (title, author, publisher, copyright date), students were asked to write a brief summary of the book; indicate the intended audience; describe how the book could be used in the classroom; comment on the book's illustrations, accuracy, and organization; and explain five science facts that were new to them, with examples from the book.

Some of the students' reviews were published in the Hardy PTA Newsletter. The student reviews will be an ongoing activity. Students will continue to read and review the new science materials, and their reviews will be published weekly in the school newsletter.

This project has had several positive results. The science book collection in the library has been updated and improved. Teachers and students are excited about the new materials and are using them more. Because teachers and students are more apt to find the information they need, they are much more satisfied. For example, with more books on science experiments, next year's science fair will be less frustrating for everyone. By reading the materials and writing reviews, students are learning critical thinking skills and improving their writing skills. In addition, other students become interested in reading some of these titles, and teachers and parents also learn about the new materials.

## High School Projects

At the high school level, librarians usually work with students and teachers to support a curriculum that is already in place. However, one librarian served as a catalyst for SMT reform by reaching out to science faculty members and working with them to develop a new eleventh-grade science course. Two other librarians enhanced the existing curriculum with projects that went beyond traditional course content.

### Computer Simulation Course Development—Sim Science

Paul VI, a Catholic high school in Fairfax, Virginia, offers, a traditional, college preparatory curriculum as well as a variety of courses for those who might not pursue postsecondary education. Librarian Diane Schule recognized the need to make science more appealing and useful to the school's students and to develop a closer working relationship between the library staff and the science faculty.

Diane's efforts paved the way for a collaboration between the librarians and the faculty of the science department that culminated in the introduction of a new eleventh-grade science course in the fall of 1995. This new course combined elements of chemistry, physics, and environmental science and used computer software, games, and simulations to involve students in problem solving and to relate science applications to life.

Initially, Diane conducted extensive research on computer-based science teaching applications and identified specific software that was available. The librarians and science teachers reviewed materials, and some demonstration programs were installed on library computers for use by selected students. After this review process, seven pieces of computer software were procured with the AAAS Institute grant. Among the topics covered were air and water pollution; elementary chemistry; the laws of motion; the Wood Car Rally, which takes a gamelike approach to introductory physics; and the Great Chemistry Knowledge Race, which allows students to make decisions and compete with each other on such topics as atomic structure, chemical bonding, and the periodic table.

This course development effort promoted greater cooperation between the science department and the library and allowed the staff to explore an alternative for teaching science.

## DNA

The Duke Ellington School of the Arts in Washington, D.C., is a magnet program designed to develop the artistic skills of students, while also providing a strong academic curriculum. The target audience for this project was the 258 tenth-, eleventh-, and twelfth-grade students studying biology, botany, and anatomy.

The objective of the project, developed by library media specialist Patricia N. Bonds, was to demonstrate how DNA is separated for analysis. A related objective was to describe the basic structure and function of the DNA and RNA molecules. DNA controls the production of protein in the cells and determines genetic makeup. Because DNA analysis has been used more and more frequently to identify criminals and to determine biological parentage, this was a topic of great interest to students.

Two teachers, the librarian, and several volunteer scientists were involved in the project. Student research on DNA began in the classroom and continued in the library. Students created charts and posters on DNA. The actual separation of the DNA took place in the classroom. Follow-up included a videotape presentation on DNA.

This project was directly related to one of the objectives required for the completion of senior high school biology. The hands-on experience enhanced the students' knowledge of DNA and made it more interesting and relevant. To support the project, the library acquired a minigel electrophoresis system for use in the experiment and a videotape titled *Bio-Chemical Basis of Biology: DNA and Protein Synthesis.*

## Internet Research Training for Science Teachers

Judy Johnson and Tamra Glover conceived and executed a very successful project in which librarians will have great interest, as many school library media specialists find themselves in the position of assisting teachers in using the Internet. These two librarians serve the needs of 160 teachers and over 2,100 students in grades 9–12 at Rapid City Central High School in Rapid City, South Dakota. All fifteen teachers in the school's science department participated in this project, a testimony to its need.

Thorough planning and preparation by the two librarians ensured the project's success. Judy and Tamra began by conducting two brief surveys of the science faculty: The first revealed that two-thirds of the science department wanted a full day of training; the second survey provided information about interest areas and the level of experience of each teacher with the Net. Data from the surveys enabled the librarians to design a customized training packet for each teacher.

The teachers prepared for the training by reviewing their customized training packets, which they received three days before the training. They then received one-on-one training with the librarians to learn search strategies. Each teacher had his or her own computer for maximum hands-on time. Once they were comfortable, the teachers could spend the rest of the day practicing and "surfing the Net." Interactive student projects were stressed, so that teachers would be able to assist students with Internet research. A list of SMT-related website addresses and activities was compiled by the librarians with input from all the participants in the training. This Website Update List is issued frequently and is shared with other libraries.

Because this training proved so successful, Judy and Tamra moved on to train other teachers in the school. The students continue to drive the teachers' interest and, synergistically, new Internet activities and projects continually surface.

## Operation Library Lab

Clara Neal, library media specialist at Dunbar Senior High School in Washington, D.C., developed a project to create an effective link between scientific learning and the library media center by expanding the center's holdings of science software. In biology, students studying anatomy and physiology dissect frogs, perch, crayfish, and earth worms. To enhance their learning, the media center purchased a CD-ROM software package that correlated with these dissection activities. To make sure that the new software was used, Clara and the biology teacher provided a staff development session for members of the science and math department. In addition, each Tuesday for two months, the library aide brought the equipment and software to the science lab and provided technical assistance for the biology teacher and students. Each Monday, the students reported to the media center to conduct research and document previous lab sessions. As part of the project, students demonstrated their work for the PTA, discussing how the use of technology, hands-on science, and cooperative education had enhanced their science learning experience.

## Quality of Drinking Water at Coolidge High School

Pauleze C. Bryant, library media specialist at Calvin Coolidge High School, NW, in Washington, D.C., formulated a project that focused on an environmental science problem that could exist in many older school buildings. Students studied the environmental and societal effects of lead as a pollutant and tested the drinking water within their school building.

Students from chemistry and social studies classes were involved in the project. The students worked in teams, studying the issue from several perspectives. The chemistry and social studies teachers, as well as the librarian, worked with the students.

Guidelines for the project were developed, and the chemistry teacher ensured that proper procedures were followed. The Environmental Protection Agency standards for PPM lead concentration in water were used, and intermittent testing was done. The social studies teacher worked with the students as they explored the societal implications of this environmental problem. The librarian coordinated the effort and ensured that students had access to the research materials they needed to pursue their investigation of the problem.

To support this project, the library purchased a lead colorimeter ($280), two refills for the colorimeter ($175), and twenty-two reverse-osmosis water bottles ($45). Exhibits from the project were displayed in the library as a culminating activity.

## Public Library Projects

The projects included in this section were developed and conducted by librarians in public libraries, often in partnership with community-based organizations.

### Birds, Birds, Birds!

Jill Randolph, children's librarian at Stow Public Library in Stow, Ohio, provided prekindergarten to third-grade students an opportunity to learn about birds and enjoy stories, poems, and factual books about them. The owner of a wild-bird store gave a short talk to the children about how to feed wild birds and how to recognize them by sight and by their calls. The children also examined and learned the purpose of flight, contour, and down feathers.

The groups listened as Jill read *Urban Roosts: Where Birds Nest in the City* by Barbara Bash and then made bird nesting boxes from wax-impregnated, recycled-cardboard kits purchased from the Phillips 66 Company.

Other books featured during the program were *Brother Eagle, Sister Sky: A Message from Chief Seattle* adapted by Susan Jeffers, *Big Friend, Little Friend: A Book about Symbiosis* by Susan Sussman and Robert James, and *Birds, Beasts, and Fishes: A Selection of Animal Poems* by Anne Carter. The children were encouraged to borrow fiction and factual books about birds following the program.

## Blast Off with Reading!

A space theme provided the motivation in this summer reading program for elementary-age children. During visits to Gary elementary schools, Mary Ann Mrozoski, children's librarian at Gary Public Library in Gary, Indiana, wore a rented space suit to stimulate children's interest and sign them up for participation. Autographed photographs of astronauts, model rockets, and books that reinforced the space theme, such as *Spaceships* and *The Space Shuttle* by Gregory Vogt and *Rockets and Satellites* by Norman S. Barrett, were displayed throughout the summer.

The space flight theme was carried through many activities during the summer. Children experimented with aerodynamics by making paper airplanes and flying saucers of different designs. Children from each age group rode on the library-sponsored float in the Fourth of July parade. Blast Off with Reading! banners, a giant model rocket, and a children's librarian in a space suit rounded out the float.

To celebrate the successful completion of the summer program, the children attended a one-hour program on exploration of the planets and stars at a local junior high school's planetarium. Prizes of balsa wood model rockets, photographs, and other educational materials, some courtesy of NASA, were awarded at the conclusion of the program.

## Build-Your-Own Tutoring Kit

The objective of this project was to give home-school parents and Indian Education tutors the experience and tools to feel confident using hands-on science activities with their children and students. Sharon M. Henry, library media specialist at Agnes Parr Resource Center, Rapid City Area Schools in Rapid City, South Dakota, and the Resource Center staff developed sixteen kits of hands-on science activities. Each activity was designed to stimulate questions that would require follow-up research by the participant. The kits included supplies, such as batteries, small bulbs, and electrical motors, for use in experiments and hands-on science activities with electricity. Tangrams were included for geometric activities. Perception, physics, and other activities were also included.

Two workshops, each one to two hours in length, were held at the Resource Center. During the workshops, eight tutors from the Office of Indian Education and seven home-school parents tried the hands-on science activities. Each participant was given a plastic tub containing the activities, supplies, and the materials for the workshop, including a booklet of worksheets and related activities. These kits were provided with AAAS Institute grant funds. One kit was retained as a model for future workshops.

These hands-on science workshops can have far-reaching effects, as each participant returns to his or her own situation to share new information and skills with a larger audience.

## Creepy-Crawly Story Time

Carolyn Caywood, area librarian at Bayside Public Library in Virginia Beach, Virginia, introduced scientific inquiry using insects and spiders as the subject. The project also encouraged children to enjoy stories, poems, and factual books about insects and spiders. The price of admission to this ongoing library program for four- through eight-year-olds and their parents was a live bug in a jar with holes poked through the lid. All the insects were displayed on a long table and admired. This was followed by a discussion of the differences among the creatures brought in by participants. To identify their insects or spiders, parents and youngsters used *Peterson's First Guide to Insects* by Christopher W. Leahy and *No Bones: A Key to Bugs and Slugs, Worms and Ticks, Spiders and Centipedes, and Other Creepy Crawlies* by Elizabeth Shephard.

Participants learned some basic characteristics of insects and enjoyed stories that reinforced the theme. Stories read at these entertaining story times included *Sam's Sandwich* by David Pelham; *Animal* by Lorna Balian; *Joyful Noise: Poems for Two Voices* by Paul Fleischman; *Someone Saw a Spider: Spider Facts and Folktales* by Shirley Climo; *Like Jake and Me* by Mavis Jukes; and *Bugs* by Nancy Winslow Parker and Joan Richards Wright.

## Earth Day

To motivate children to read stories and factual books about ecology and the environment as part of Earth Day celebrations, Caroline Parr, children's librarian and coordinator of children's services at the Central Rappahannock Library in Fredericksburg, Virginia, led discussions about recycling and preserving the environment with boys and girls from first to fifth grade. The children listened to stories related to the theme, such as *The Little House* by Virginia Lee Burton and *Where the Forest Meets the Sea* by Jeannie Baker. The children planted bean seeds in paper cups and created toys from recycled materials. At the end of the program, each child pulled a slip of paper describing a simple recycling or conservation activity from a globelike bowl and resolved to begin that practice. Ideas for activities were drawn from *Recylopedia* by Robin Simons and *50 Simple Things Kids Can Do to Save the Earth* by EarthWorks Group.

The librarians created an attractive four-page brochure on bright, standard-size paper, featuring a picture of Earth on the cover. The brochure included two pages of recommended titles under the following headings: The Natural Environment, The Threatened Environment, Endangered Species, Overcrowding, Recycling, and Things to Do. The title, author, reading level, and call number were listed for each book. On the last page of the brochure were several other subject headings that library users could explore to find more materials related to the ecology and recycling.

## The Effects of Weather on the Air Force Mission

Most of the children from the military families at Ellsworth Air Force Base (AFB) in South Dakota attend school in the Douglas Public School System. Approximately 60 percent of the 1,800 students at the Douglas Elementary and Middle Schools are military dependents. These children, along with the children of base employees, also use the Holbrook Library, located on the base. The goal of this project was to encourage these students to seek information at the base library, especially for SMT assignments, projects, and questions. Barbara Misselt, library technician, designed a project to show the impact of weather on the ability of the pilots to accomplish their mission—flying the B-1 bomber—and to show the weather-related resources available at the library.

A great deal of cooperation between the AFB community and the school district was necessary, and many individuals volunteered their time to ensure the success of the project. The target audience was the eighth graders at Douglas Middle School in Box Elder. The two eighth-grade science teachers, Kelly Lane and Tony Burns, readily embraced the idea. Capt. George Smith, a B-1 pilot who is also an avid library user, agreed to make a presentation to the students. The weather station provided Doppler and satellite weather maps. The science teachers prepared overhead transparencies of the photos that Capt. Smith wanted to show the students, and the school librarian brought in a display of books about weather.

Barbara and Capt. Smith spent a day at Douglas Middle School, making five presentations to over 200 attentive students. Barbara spoke briefly about how the students could use the base library to do SMT-related research. Capt. Smith talked about the importance of learning as much as possible from school and from libraries. He shared some of his flying experiences, stressing the importance of preflight weather briefings and coordinated information from weather stations all along the path of a long flight. He described how, on a long flight, the B-1 must refuel in the air, supported by KC-135 tankers all along its route. For midair refueling to be successful, these flights must have favorable weather conditions.

Barbara gave each student a handout inviting them to the library and listing Internet sites for weather exploration. Right after school that day, three youngsters were in the base library trying out the weather addresses on the Internet. Student use of the base library has increased, as has Internet usage in the Douglas Middle School library and classrooms.

## Go Green at the Library

The Rhode Island Department of Library Services and children's librarians throughout the state developed and participated in a summer reading program for elementary-age children designed around an environmental science theme. Children read books about protecting the environment, attended an awareness-raising program presented by a paid performer, and participated in an antilitter activity. The antilitter activities ranged from distributing homemade litterbags, to planting an organic garden, to participating in a neighborhood cleanup day.

Green Pickle Storytimes were targeted to kindergartners through second graders and featured stories with a nature theme and ecologically sound crafts. During one program, the children listened to *Jack and the Beanstalk* and then planted their own bean seeds in recycled cottage cheese containers.

Fourth through sixth graders were recruited to clean up the library grounds and plant and maintain flowers throughout the summer. The librarians used books on plants to help the youngsters learn which plants would thrive in different locations around the libraries. The children also read books, including *The Garden of Abdul Gasazzi* by Chris Van Allsburg, *Keepers of the Earth: Native American Stories and Environmental Activities for Children* by Michael J. Caduto and Joseph Bruchac, and *What on Earth Can You Do with Kids: Environmental Activities for Every Day of the School Year* by Robin Freedman Spizman and Marianne Daniels Garber.

## How to Make a Science Fair Project!

Mary Alice Deveny, youth librarian at Selby Public Library in Sarasota, Florida, designed and conducted this project to help students get started on their science fair projects and become familiar with the science resources available to them. The Sarasota County School System's science fair director presented information about developing a science fair project for elementary and middle school students and their parents. The one-hour presentation covered the following areas: selecting a topic; using scientific methods to develop the project; following the guidelines of the school system; keeping a logbook to record data and observations; and creating an attractive and effective project board. The talk was followed by a question-and-answer session.

Handouts at the presentation recommended books that contained resources for science fair projects. Children and parents also received *Exploring Nature: Field Trips for Families,* a guide to twenty-eight outings from Tampa to Naples; and "A Science Fair Project Pathfinder," prepared by Mary Alice and other children's librarians (see the section of the same name later in this chapter).

## Kitchen Chemistry

This public library Science in the Summer program for six- to eleven-year-olds was intended to inspire the children to explore chemistry through stories and factual books. Nancy Gifford, children's librarian at the Schenectady County Public Library in Schenectady, New York, and an avid cook, explained to the boys and girls the leavening properties of baking powder, yeast, and baking soda with vinegar. The group observed the chemical reactions created when sugar is added to warm water and yeast and marveled at the volcano-like eruption of baking soda combined with vinegar. The children tasted and compared biscuits made with and without baking powder and salty and salt-free potato chips. Nancy read *The Lady Who Put Salt in Her Coffee* adapted by Amy Schwartz, and *Salt: A Russian Folktale* by Jane Langton.

A book display and book list were available for children who wanted to explore chemistry on their own. Resources for the program were *Science Experiments You Can Eat, More Science Experiments You Can Eat,* and *Chemically Active!: Experiments You Can Do at Home,* all written by Vicki Cobb.

The Science in the Summer program for children at the Schenectady County Public Library was made possible in part through funding from the ELFUN Society of General Electric.

## Lively Ladybugs

Phylis Franklin, children's service coordinator at Margaret Cooper Public Library in Linton, Indiana, developed a program to provide preschoolers with an opportunity to observe live ladybugs and enjoy stories and factual books about them. The program began with the librarian reading Eric Carle's *The Grouchy Ladybug* and *The Life Cycle of a Ladybug.* She then brought out a glass jar of live ladybugs for the children to observe. The boys and girls learned that ladybugs have six legs and three body sections and are a boon to gardeners because they eat garden pests like aphids.

The children each made a ladybug out of two 8-inch circles of construction paper, one black and one red. The red circle was cut in half and glued to the black circle with corners touching at one end. This gave the appearance of a ladybug spreading her wings. The children used black crayons to make spots on the wings. *The Ladybug and Other Insects* by Pascale de Bourgoing and a number of other books were displayed and loaned.

## NatureConnections

Jane Sorenson, project director and children's librarian for the Chicago Public Library System, Chicago, Illinois, sought to interest elementary-age children in library materials about natural history and to establish a link between the library collection and the many educational institutions in Chicago. Through a grant from the Hermon Dunlap and Ellen Thorne Smith Fund of the Chicago Community Trust, NatureConnections collaborated with a network of zoos, museums, and science centers in Chicago. Project materials, housed at forty-three Chicago Public

Library locations, include a unique collection of natural history books, magazines, videos, pamphlets, and realia.

The librarians work with representatives of the various institutions to select the best reading materials to supplement their exhibits. The museums and nature centers loan display materials for exhibit in the NatureConnections locations. The library sponsors staff development science-related field trips, meetings, and workshops for the children's librarians.

NatureConnections has developed materials that enhance the connection between natural history materials at the library and the zoos, museums, and nature centers in Chicago. On one game sheet, for example, children are asked to match Chicago landmarks with an appropriate children's book: Shedd Aquarium matches *One Fish Two Fish Red Fish Blue Fish* by Dr. Seuss, the Adler Planetarium matches *Goodnight Moon* by Margaret Wise Brown, and *Danny and the Dinosaur* by Syd Hoff matches the Field Museum of Natural History.

## Portable Pockets

Karen Slachta, children's librarian at Interboro Unified Districts Library in Peckville, Pennsylvania, wanted to provide an opportunity for preschoolers to observe nature and to enjoy stories and factual books about nature. During the fall, a children's librarian read stories about nature to the children and talked about what they could find outside, such as colorful leaves, acorns, rocks, and pine cones. The children decorated paper lunch bags to use as portable pockets during a nature walk that they would take with a parent sometime during the week.

The following week, the boys and girls brought their portable pockets back to the library, excited to show off their discoveries. The story time centered on discussing subjects stimulated by their curiosity about the objects they had found. Some books recommended for the program were *Peter's Pockets* by Eve Rice, *Fall* by Ron Hirschi, *The Berenstain Bears' Nature Book* by Stan and Jan Berenstain, and *A Pocketful of Cricket* by Rebecca Caudill.

## READiscover Planet Earth

Lynda Welborn, Senior Consultant/School Library Media at the Colorado State Department of Education, and children's librarians throughout the state developed and promoted a summer reading program for elementary-age children with an environmental science theme. The boys and girls began learning about landfills from *Trash?* by Charlotte Wilcox. They conducted a decomposition experiment by burying pieces of vegetables, newspapers, and styrofoam in separate piles in the library yard and watering the soil. They made labels indicating each item and the date it was buried. During subsequent weeks, the children dug up the items, compared the stages of decomposition, and recorded their observations in a journal before reburying them.

The children also made collages with environmental themes by cutting out pictures, words, and letters from junk mail. They created costumes from recycled objects, such as dust mops, pillows, boxes, bags, and stockings.

*Amy's Dinosaur* by Syd Hoff, *The Salamander Room* by Anne Mazer, *The Mountain* by Peter Parnall, *The Talking Earth* by Jean Craighead George, *Tunafish Sandwiches* by Patty Wolcott, and *Just a Dream* by Chris Van Allsburg were a few of the stories read during these programs.

## Rocket Launch

Catherine Christman, children's librarian at Charleston Public Library in Charleston, South Carolina, devised a program for six- through twelve-year-olds focused on rocketry and space travel. Cadets from a nearby Air Force ROTC presented a one-hour program on rocketry. The cadets explained to the children some principles of aerodynamics and how a three-stage rocket works. The program culminated with the launch of a solid fuel, three-stage model rocket in a park adjoining the library grounds. The children watched in delight as the rocket blasted off, then separated into stages before floating back to earth on parachutes.

A collection of books, such as *Rockets and Satellites* by Franklyn M. Branley, *To Space and Back* by Sally Ride and Susan Okie, and *Easy-to-Make Projects in Space Science* by Robert Gardner were displayed for browsing and borrowing by the children.

## A Science Fair Project Pathfinder

This project created a brochure for elementary and middle school students and their parents that described science fair resources available to them. Mary Alice Deveny and other children's librarians at Selby Public

Library in Sarasota, Florida, produced *Science Fair Project Pathfinder,* an attractive, four-page brochure on bright yellow paper. The brochure was distributed throughout the school year, saving the staff from repeating the same information hundreds of times and providing concrete guidance for the students and their parents.

The *Pathfinder* featured two maps, one of which showed the interior of the library with the children's library, the periodicals section, and the adult reference area circled. The other map provided directions to the Ann Marbut Environmental Library, a special collection open to the public and located a few miles away.

The *Pathfinder* encouraged amateur scientists to browse the stacks by listing the 500s and 600s of the Dewey Decimal Classification, both broken down to ten subject headings. Short lists of reference sources, organized by location on the corresponding map, included call numbers and a space for a check mark after the book or resource had been examined. The reference sources were divided into separate locations: juvenile reference, adult reference, and the environmental library. In addition, other resources in the community, such as school and university libraries, businesses, hospitals, museums, and government agencies, were listed as sources for further research.

## The Think Tank

The Children's Services department of the Stow Public Library in Stow, Ohio, headed by Carolyn Morgan Burrier, used a $10,000 bequest to create the Think Tank, a science, math, and geography area filled with interactive materials. The Think Tank featured an Apple computer loaded with science, math, and thinking skills software, such as *Where in the World Is Carmen Sandiego?, Where in Time Is Carmen Sandiego?,* and *The Tree House.* Microscopes, puzzles, prisms, magnets, kaleidoscopes, plants, and rabbits encouraged the youngsters to observe, experiment, make discoveries, and read. Displays were frequently added to or changed, providing children with the opportunity to do such things as grow crystals; learn how to run a clock with a potato; and observe, measure, and record the growth of an amaryllis.

Many of the ideas were based on hands-on activities and experiments described in science books in the collection, including *Science Fun with a Homemade Chemistry Set* by Rose Wyler, *The Science Book of Sound* by Neil Ardley, and *Fun with Science: Light* by Brenda Walpole.

## Thomas Edison

As part of a public library program for families, the Indiana Humanities Council subsidized performances of a one-man play about Thomas Edison written and acted by an Indiana scholar-artist. The career and personal life of Edison were revealed to the audience through the personas of Edison, his son, Henry Ford, and a stagehand. The program focused on the myths surrounding Edison, his place in history, and the personal costs of genius.

The Gary Public Library in Gary, Indiana, hosted one of these performances, which was open to the public. Families were especially encouraged to attend. Mary Ann Mrozoski, the children's librarian, created a display of children's books about Edison and his inventions and about other scientists, inventors, and the invention of common household objects. *Thomas Alva Edison: Great Inventor* by David A. Adler, *Samuel Todd's Book of Great Inventions* by E. L. Konigsburg, and *Mistakes That Worked* by Charlotte Foltz Jones were among the books displayed in the library for browsing and borrowing before and after the performance.

# APPENDIX
## About the AAAS Science Library Institute

The AAAS Science Library Institute was a professional development program funded by the U.S. Department of Education. The first AAAS Science Library Institute project was conducted in Washington, D.C., in 1993–1994. The second was conducted in Rapid City, South Dakota, in 1995–1996. Both projects sought to train librarians to assume key roles in providing science, mathematics, and technology (SMT) learning opportunities for all students as well as to make libraries a focal point of SMT reform.

The Institute combined theoretical training in SMT reform issues with practical training in how to use hands-on science and mathematics activities in the library. The five workshops focused not only on traditional library skills and content, such as collection development and evaluation, but also on planning and conducting inquiry-based science and mathematics programs. Participants attended five daylong sessions, plus an additional workshop on how to use the Internet. After the workshops, assisted by Institute mini-grants, participants used the skills they had developed to conduct science programs in their libraries. The goals of the Institute were to:

enhance the skills of librarians in selecting accurate and effective science resources that support educational reform and reflect equity issues that address the needs of minority and disadvantaged youth;

increase the number of librarians who can play a key role in science and mathematics reform by providing them with access to a matrix of concepts and topics recommended by current national standards projects, such as AAAS's Project 2061 and the *National Science Education Standards;*

increase in-school and out-of-school science learning opportunities for children by providing hands-on science training and support to school and public librarians so that they can act as in-service workshop leaders in their libraries and school systems;

create partnerships between public and school librarians and members of the scientific community who will assist in reviewing collections and incorporating scientifically accurate materials as well as in planning science opportunities for children; and

train librarians to become disseminators of new science and mathematics education projects, products, techniques, and exemplary practices to science teachers, parents, and community members.

The AAAS Science Library Institute addressed the need for librarians both to hone their skills in selecting library resources and to develop partnerships in the community that ensure continued support for those resources. Some of the specific topics and skills addressed by the Institute are described here.

In the Institute workshops, AAAS staff and other presenters articulated the components of a good science book and asked participants to assess their own science collections with regard to these criteria. Most participants expressed concern about the age of their collections, but were reluctant to weed out some of the older books because their science collections were so small. In entry surveys, participants also noted that their confidence in selecting science materials was not as great as their confidence in selecting language arts or social studies resources.

Participants were given a list of equity guidelines and asked to consider whether the resources they were reviewing for scientific accuracy also reflected our diverse, multicultural society. Participants discussed the importance of omitting stereotypical portrayals of minority groups and scientists from children's science materials, and then examined their own resources for examples of these portrayals.

In the workshops, participants practiced their resource selection skills in a variety of ways. In addition to reviewing books, participants were given the opportunity to enhance their science collections through the mini-grant program. They selected a wide variety of resources, ranging from books to videos to software. Finally, participants received bibliographies and other sources for identifying exemplary science and mathematics materials.

Participants also received copies of *Science for All Americans, Benchmarks for Science Literacy, Great Explorations,* and, in the Rapid City program, the K–12 Science Standards for South Dakota. After reading *Science for All Americans,* librarian Barbara Allum wrote the following in her project journal: "I understand that we are not going to have an all-encompassing solution to this problem, but if I as a librarian can help influence a child's life—I'm ready to start. I knew our students' test scores in science and math were down, but I didn't know that I'd have much to do with the solution."

A key benefit to librarians was an increased ability to provide informal science activities in the library. These activities were not intended to replace science instruction in the classroom, but rather to augment opportunities for children to explore and experience science. Many librarians selected projects with a hands-on science component as part of their mini-grant activities. (For descriptions of many of these projects, see chapter 9.) Reports on these projects indicated that the librarians worked closely with science teachers to ensure that supplemental activities in the library supported ongoing instructional objectives. The training in science and math reform issues and pedagogy that the librarians received facilitated increased collaboration with classroom science teachers.

During the workshops, librarians were given an assignment to review a hands-on science book from their collection and to describe how they might set up a library display based on an activity in the book. Librarian Jane Holeton selected *Colorful Light* by Julian Rowe and Molly Perham. She developed an activity on color, which she conducted with a variety of students of different grade levels. This is how she described her reaction to the experience: "Since I have a limited science background, I was a little worried about doing this activity, because I felt I wouldn't be able to answer questions. But I read in a magazine that you don't need to know all the answers, because actual scientists don't know all the answers. So that made me feel a little more at ease. But after doing this activity and seeing the excitement and learning taking place, I know that I will be doing many more of the hands-on science activities."

Librarians participating in the Institute were each given $500 and asked to devise an action plan for a project they would implement in their library or school. They conceptualized and developed many exciting projects, including a six-week aviation project and super-plane contest; a three-school health fair; a Science Everywhere Fair, in which middle-school children designed hands-on projects for elementary-school students; and a project to raise students' awareness about geological features in the area in which they live—the Black Hills. (See chapter 9 for more examples.)

Each Institute participant created a poster depicting her or his project. In the Washington, D.C., project, the posters were displayed at the Martin Luther King Jr. Branch of the D.C. Public Library. A similar activity took place in Rapid City where posters were displayed during Library Day at a Rapid City shopping mall.

Participants completed evaluation questionnaires at the beginning and end of each session. They also kept a journal of observations and experiences as a "homework" assignment. Finally, each submitted a report that described the outcomes of her or his project and compared the results to the goals outlined in the action plan. These reports, along with follow-up phone calls by the project evaluator, Dr. Mary Chobot, were used to prepare final reports on the impact of each Institute.

Workshop features that participants found particularly useful were the sharing of project ideas; information about resources available in the community; information about and practice in writing grants; hands-on science training; networking with other librarians; planning, justifying, and conducting a project; and building leadership skills. The librarians also responded positively to the opportunity to brainstorm with peers on ways in which they could successfully use science in their libraries.

Many of the Institute participants' feelings can be summarized in the following statement by Mary Pfeifle: "The benefits obtained from working on a hands-on project are greater rapport with students, promotion of enthusiasm for science and science careers, better public relations within the school and community, increased visibility of the library program within the school and community, and the personal rewards of providing a fun learning experience for our students."

The AAAS Science Library Institute sought to develop participants' abilities to conduct in-service workshops on equity and excellence in science and mathematics education. Librarians were able to use their workshop training to become key players in educational reform in their schools. This result can be replicated in local communities either through in-service workshops or through less-formal gatherings of librarians with similar goals.

We recognize that it is not possible to provide a science or mathematics background for every librarian. However, as trained information specialists, librarians can learn to access and use tools already at their disposal to help make informed decisions about materials and programs for their libraries. They can also provide teachers and parents with information about innovative and effective strategies for improving students' science and mathematics performance. Well-informed educators, parents, and community leaders increase the likelihood of meaningful and effective school reform. We believe that any community—urban, suburban, or rural—can benefit from a Science Library Institute. We also believe that all librarians can become leaders in the effort to expand science and mathematics literacy for all children.

# CONTRIBUTORS

**Jerry A. Bell** is a senior scientist with the Educational and International Activities division of the American Chemical Society. He previously served as a program director for the AAAS. Bell is a former college and university faculty member and National Science Foundation (NSF) division director in science education. He received his Ph.D. in chemistry from Harvard University in 1962. He is the author, co-author, or editor of many publications in chemical dynamics research and science education, especially hands-on, inquiry-oriented activities.

**Mary C. Chobot** received a B.A. in English from Le Moyne College, an M.S.L.S. in library science from Syracuse University, and a Ph.D. in educational psychology and evaluation from Catholic University of America. She was a project consultant and evaluator for the AAAS Science Library Institute Projects in Washington, D.C. and in South Dakota. She was also a consultant to the Library of Congress on the American Memory Project. Chobot has taught library and information science at the college level at Catholic University and at Syracuse University.

**Tracy Gath** is now the editor of SB&F Online, she served as editor of *SB&F* for three years. She is the co-editor of *SB&F's Best Books for Children 1992–95,* provides other editing services for AAAS's *Directorate for Education and Human Resources Programs,* and wrote for and edited publications for Science and Literacy for Health projects, including the literacy toolkit "How Drugs Affect the Brain," as well as *The Brain Book* and *Brain and Behavior.* Gath received her bachelor's degree in environmental science from Antioch College in Yellow Springs, Ohio. She served on the board of directors for the Academy of Hope, an adult education program in Washington, D.C.

**Barbara Holton** works for the federal government in the field of library science and was a public librarian in suburban Maryland. She earned an M.S. in library science from the Catholic University of America in Washington, D.C., and an M.A. in measurement, statistics, and evaluation from the University of Maryland.

Her experience with and interest in science education includes teaching science to fifth and sixth graders, as well as presenting hands-on science programs to children. While a library school student, she was involved in an independent study to develop evaluation criteria for science programs for children, write science program descriptions, research resources for science programs and present a collection of science programs for children at several library conferences in Virginia and Maryland. As a participant in the AAAS Science Library Institute, Holton learned more about science education and science concepts through presentations made by scientists in various fields and through inquiry-oriented activities.

**Coleen Salley** is a Professor Emerita of Children's Literature at the University of New Orleans. She has traveled the United States, Europe, and Asia as a storyteller and is a frequent presenter at state, regional, and national conferences of reading teachers, science teachers, and librarians.

**Maria Sosa** is the editor-in-chief of *SB&F,* manages the publications staff of the *Directorate for Education and Human Resources Programs* at AAAS, and is also the project director for the *Science and Literacy for Health Project.* Sosa received her master's degree in education and the teaching of English from Teachers College, Columbia University, and a bachelor's degree in humanities from Shimer College. Before coming to AAAS, she was an award-winning software and audiovisual writer/ producer, as well as a writer and editor of educational print materials. Since coming to AAAS, she has conducted a number of workshops for librarians at national meetings and represents AAAS at national and local meetings of community-based organizations and professional scientific associations. She is also the editor of a monograph on children's science books.

**Terrence E. Young** is a school library media specialist at West Jefferson High School is Harvey, Louisiana. His articles have appeared in *SB&F, Knowledge Quest, Library Talk,* and *Book Report.* He is a frequent presenter at conferences for teachers and librarians.

# INDEX